UK Environmental Policy in the 1990s

Edited by

Tim S. Gray

Senior Lecturer and Head of the Department of Politics
University of Newcastle

First published in Great Britain 1995 by
MACMILLAN PRESS LTD
Houndmills, Basingstoke, Hampshire RG21 6XS
and London
Companies and representatives
throughout the world

A catalogue record for this book is available
from the British Library.

ISBN 0–333–62120–4 hardcover
ISBN 0–333–62121–2 paperback

First published in the United States of America 1995 by
ST. MARTIN'S PRESS, INC.,
Scholarly and Reference Division,
175 Fifth Avenue,
New York, N.Y. 10010

ISBN 0–312–12672–7

Library of Congress Cataloging-in-Publication Data
UK environmental policy in the 1990s / edited by Tim S. Gray.
p.   cm.
Includes bibliographical references and index.
ISBN 0–312–12672–7 (cloth)
1. Environmental policy—Great Britain.   2. Great Britain-
-Environmental conditions.   I. Gray, Tim S. 1942–
GE190.G7U3   1995
363.7'00941'09049—dc20   95–5580
                                                          CIP

10   9   8   7   6   5   4   3   2   1
04   03   02   01   00   99   98   97   96   95

Printed and bound in Great Britain by
Antony Rowe Ltd, Chippenham, Wiltshire

I would like to dedicate this book to **Mark, Guy** and **Claire**, whose future quality of life will be significantly affected by the success or failure of the UK government's environmental policies

Contents

Notes on the Contributors   ix

Acknowledgements   xiv

List of Abbreviations   xv

1 Introduction   1
   Tim S. Gray

2 View from the Inside: UK Environmental Policy
   Seen from a Practitioner's Perspective   11
   Tom Burke

3 Impact of the European Union on UK Environmental
   Policy Making   18
   Nigel Haigh and Chris Lanigan

4 The Establishment of a Cross-Sector Environment Agency   38
   Neil Carter and Philip Lowe

5 The Precautionary Principle in UK Environmental
   Law and Policy   57
   Andrew Jordan and Timothy O'Riordan

6 UK Environmental Policy and the Politics of
   the Environment in Northern Ireland in the 1990s   85
   Steven Yearley

7 The Politics of Mutual Attraction? UK Local
   Authorities and the Europeanisation of
   Environmental Policy   101
   Stephen Ward

8 The UK and Global Warming Policy   123
   David Maddison and David Pearce

9 Energy Conservation Policy   144
   Gerald Manners

10   Constructing Regulations and Regulating
     Construction: The Practicalities of Environmental Policy   159
     Elizabeth Shove

11   UK Environmental Policy and Transport   173
     Kenneth Button

12   Acid Rain: A Business-as-Usual Scenario   189
     Jim Skea

13   Nuclear Waste Disposal: A Technical Problem in
     Search of a Political Solution   210
     Andrew Blowers

14   Running Up the Down Escalator: Developments in British
     Wildlife Policies after Mrs Thatcher's 1988 Speeches   237
     Stephen C. Young

15   The Tragedy of the Common Fisheries Policy: UK
     Fisheries Policy in the 1990s   263
     Anthony Stenson and Tim S. Gray

16   The UK and the International Environmental
     Agenda: Rio and After   283
     Michael Redclift

Index   303

Notes on the Contributors

Andrew Blowers is Professor of Social Sciences (Planning) at the Open University, a member of the Radioactive Waste Management Advisory Committee, and has published extensively on environmental politics (especially the politics of pollution and radioactive waste), planning theory and politics and local government and politics. He has co-authored, edited or co-edited scores of publications, including *Something in the Air: Corporate Power and the Environment* (1984), *Nuclear Power in Crisis* (1987), *The International Politics of Nuclear Waste* (1991), and *Planning for a Sustainable Environment* (1993).

Tom Burke is Special Adviser to the Secretary of State for the Environment. After several years with Friends of the Earth in the 1970s, becoming its Executive Director, Tom Burke became Director of the Green Alliance from 1982 to 1993 (an environmental think-tank). In 1987, together with Julia Hailes and John Elkington, he founded Sustainability, a private consultancy company, set up to promote environmentally sustainable growth. He has written and broadcast extensively on environmental matters and is co-author of *The Green Capitalists* (1987), *Green Pages* (1988), and *Ethics, Environment and the Company* (1990).

Kenneth Button is Professor of Applied Economics and Transport and Director of the Centre for Research in European Economics and Finance at Loughborough University, and Visiting Professor in Transport and the Environment at the Tinbergen Institute, Amsterdam. He has held visiting academic posts at the University of California, Berkeley, and the University of British Columbia, and he has been a full-time consultant to the Organisation for Economic Cooperation and Development (OECD) in Paris. In 1994 he was Special Adviser to the House of Commons Transport Committee. He has published extensively in the area of transport policy.

Neil Carter is Lecturer in Politics, University of York, convenor of the Political Studies Association specialist group on environmental politics and a member of the Editorial Board of *Environmental Politics*. He is co-author of *How Organisations Measure Success* and author of many articles and chapters on green issues, and is currently writing a book with Andrew Flynn on *The Politics of the Environment*.

Tim S. Gray is Senior Lecturer and Head of the Politics Department, University of Newcastle upon Tyne. Trained as a political theorist, his authored and co-authored publications include books and articles on freedom, Edmund Burke, Flora Tristan and Herbert Spencer. Now moving into environmental theory and practice, his current research is in green theory (including Gaia, Rousseau and environmental justice); the International Whaling Commission; UK fishing policy; and British environmental policy as a whole.

Nigel Haigh, OBE, is Director of the Institute for European Environmental Policy, London (IEEP), which is an independent body for the analysis of environmental policies in Europe. It is linked to similar bodies in other countries. He is Visiting Fellow at Imperial College Centre for Environmental Technology and Honorary Research Fellow at the Faculty of Laws, University College, London. He has published extensively on environmental policy matters including the *Manual of Environmental Policy: The EC and Britain* (1993).

Andrew Jordan is Research Associate at the Centre for Social and Economic Research on the Global Environment (CSERGE), located jointly at the University of East Anglia (UEA) and University College London (UCL). He has published many articles and chapters on environmental policy, and is currently engaged on a long term study of how governments, business and other agencies perceive the threat of global environmental change, and how they are responding or adapting to it.

Chris Lanigan is Ridley Fellow, Politics Department, University of Newcastle upon Tyne, and a doctoral student researching into the regional identity of the North-East of England. His master's degree focused on the politics of the European Union.

Philip Lowe has been Duke of Northumberland Professor of Rural Economy since 1991 in the Department of Agricultural Economics and Food Marketing of the University of Newcastle, and is also Director of the Centre for Rural Economy, and Co-Director of the Economic and Social Research Council (ESRC) Countryside Change Initiative. He has authored, co-authored or edited over 20 books, 30 chapters and 40 articles on environmental and agricultural policy and politics. His most recent co-authored books are *The Green Wave: Ecological Parties in Global Perspective* (1995) and *Constructing the Countryside* (1993), and he has co-edited *European Integration and Environmental Policy* (1993).

David Maddison is Research Fellow at the CSERGE at UCL, and also a PhD student in the Economics Department of Strathclyde University. His research interests include methodological aspects of studies seeking to evaluate the macro-economic impact of environmental policies and cost-benefit analyses of slowing climate change.

Gerald Manners is Professor of Geography, UCL, and sometime Visiting Fellow at the Center for Resources for the Future, Washington, DC, and at Harvard University's, Massachusetts Institute of Technology, and the Australian National University. He is currently Adviser to the House of Lords Select Committee on Sustainable Development and Consultant to the Association for the Conservation of Energy (ACE), and was previously Specialist Adviser to the House of Commons Select Committee on Energy. He has published widely on energy issues; his books include *Coal in Britain: An Uncertain Future* (1981).

Timothy O'Riordan is Professor at the School of Environmental Sciences at UEA and Associate Director of CSERGE. He has authored, co-authored and edited scores of books, chapters and journal articles on environmental politics, environmental policy, countryside management and environmentalism. He is currently editing a monograph on the precautionary principle in environmental management.

David Pearce is Professor of Environmental Economics at UCL and Director of CSERGE. He is the author, co-author and editor of some 200 articles and over 30 books, including *Blueprint for a Green Economy* (1989), *Blueprint 2* (1991), *Blueprint 3* (1993), *World Without End* (1993), and *Economics of Natural Resources and The Environment* (1991). His research interests are entirely in the area of environmental economics. He is an adviser to the Vice President of the World Bank and to the Administrator of the United Nations Development Fund, and he has been a Personal Adviser to two UK Secretaries of State for the Environment. He holds the United Nations Global 500 Award for services to the world environment.

Michael Redclift is Professor of Rural Sociology, University of London, and research coordinator of the ESRC Global Environmental Change Programme (which is the biggest of its kind in the UK, and is scheduled to last until 1998 with a budget of over £20 million). His authored, co-authored and co-edited publications include six books on agriculture, the environment and rural development, including *Sustainable Development:*

Exploring the Contradictions (1987), and over 80 journal articles and book chapters. His research interests focus particularly on sustainable resource use in developing countries and in Europe, and global environmental change.

Elizabeth Shove is Senior Lecturer in the School of Social and International Studies at the University of Sunderland. Although a sociologist, she has much experience of government funded building research having previously been employed as Research Fellow at the Institute of Advanced Architectural Studies at the University of York. Participation in the ESRC's Global Environmental Change Programme has created further opportunities for her to pursue her research interests in the sociology of building and the environment and in the social organisation of technical change.

Jim Skea is Senior Research Fellow, Science Policy Research Unit (SPRU), University of Sussex and leader of its Environment Programme. He is author and co-author of many publications on energy technology and policy, integration of environmental concerns to industry, and pollution in Britain and Europe, including *Acid Politics: Environmental and Energy Policies in Britain and Germany* (1991), with S. Boehmer-Christiansen.

Anthony Stenson is a doctoral student, funded by the British Academy, in the Politics Department of the University of Newcastle upon Tyne. His PhD research is on the place of technology in green political theory. His study of UK fishing policy has been co-funded by the Research Committee and the Politics Department of Newcastle University.

Stephen Ward is a Research Associate in Environmental Politics at the Centre for Rural Economy, University of Newcastle upon Tyne. He is completing his PhD on environmental agendas in UK local government, and is currently working on a research project on the effects of the EU on British environmental politics and administration. He has published several articles on local authorities and environmental policy.

Steven Yearley was recently appointed to a Chair in Sociology at Queen's University of Belfast, having previously been Professor of Sociology at the University of Ulster. He has published extensively on the sociology of environmental organisations, on the role of science in environmental controversies, and on environmental issues in Northern Ireland. His most recent book is *The Green Case* (1992).

Stephen C. Young is Senior Lecturer at Manchester University, joint editor of *Environmental Politics* and editor of the Newsletter of the European Consortium for Political Research's Standing Group on Green Politics. Author of *The Politics of the Environment* (1993) and co-author of *Cities in the 1990s* (1993), he has published several articles and chapters on wildlife and sustainable development issues. He has an ESRC award for a project on the consequences of the Rio Earth Summit for local government in Britain, Holland and Denmark.

Acknowledgements

I would like to acknowledge the financial help given to meet the expenses of the Colloquium held in December 1993 in the Politics Department of the University of Newcastle, the papers for which form the basis of this edited book. Funding was obtained from three sources: the ESRC's Global Environmental Change Programme's Networking and Dissemination Activities Fund; the Research Committee of the University of Newcastle upon Tyne; and the block grant of the Department of Politics.

I am pleased also to acknowledge the encouragement given to me by the contributors to the Colloquium, and in particular the generous support of Professor Philip Lowe. I am grateful to my graduate research students, Pam Finlay, Hyoungwook Jeong, Mike Bray and Harriet Mitchell, for their help with the Colloquium arrangements.

Finally, I must thank my secretary, Margaret Hill, both for her heroic transcription of the Colloquium proceedings, and for her expert preparation of the manuscript for the publishers.

TIM S. GRAY

List of Abbreviations

ACC	Association of County Councils
ACE	Association for the Conservation of Energy
ADAS	Agriculture Development Advisory Service
ADC	Association of District Councils
AGR	Advanced Gas-cooled Reactor
ALARA	as low as reasonably achievable
AMA	Association of Metropolitan Authorities
AONBs	Areas of Outstanding Natural Beauty
ASSIs	Areas of Special Scientific Interest
BAT	best available (abatement) technology/techniques
BATNEEC	best available techniques not entailing excessive costs
BBA	British Board of Agrément
BNFL	British Nuclear Fuels Limited
BPEO	best practicable environmental option
BTCV	British Trust for Conservation Volunteers
CAP	Common Agricultural Policy
CBI	Confederation of British Industry
CCGT	combined cycle gas turbine
CCW	Countryside Council for Wales
CEGB	Central Electricity Generating Board
CEMR	Council for European Municipalities and Regions
CFCs	chlorofluorocarbons
CFP	Common Fisheries Policy
CHP	Combined Heat and Power
CLA	Country Landowners Association
CLEAR	Campaign for Lead Free Air
CNCC	Council for Nature Conservation and the Countryside
CND	Campaign for Nuclear Disarmament
COMARE	Committee on the Medical Aspects of Radiation in the Environment
CPRE	Council for the Protection of Rural England
CSERGE	Centre for Social and Economic Research on the Global Environment
CWTs	County Wildlife Trusts
DED	Department of Economic Development
DoE	Department of the Environment
DoE NI	Department of the Environment in Northern Ireland

DoT	Department of Transport
DTI	Department of Trade and Industry
EA	Environment Agency
EC	European Community
ECJ	European Court of Justice
EDG	Environment Dialogue Group
EEO	Energy Efficiency Office
EHOs	Environmental Health Officers
EIA	environmental impact assessment
EIS	European Information Service
EMAS	Energy Management Assistance Scheme
EN	English Nature
ENDS	Environmental Data Services
EPA	Environmental Protection Act 1990
ERDF	European Regional Development Fund
ESAs	Environmentally Sensitive Areas
ESRC	Economic and Social Research Council
EST	Energy Saving Trust
EU	European Union
FGD	flue gas desulphurisation
GATT	General Agreement on Tariffs and Trade
GDP	gross domestic product
GEF	Global Environment Facility
GESAMP	Group of Experts on the Scientific Aspects of Marine Pollution
GHGs	Greenhouse gases
GMOs	genetically modified organisms
GNP	gross national product
GWPs	global warming potentials
HEES	Home Energy Efficiency Scheme
HLW	high level waste
HMIP	Her Majesty's Inspectorate of Pollution
IAEA	International Atomic Energy Agency
IBRD	International Bank for Reconstruction and Development
ICRP	International Commission for Radiological Protection
IEEP	Institute for European Environmental Policy
IEHO	Institute of Environmental Health Officers
IGEE	Inter-departmental Group on Environmental Economics
ILW	intermediate level waste
IMF	International Monetary Fund
IPC	integrated pollution control
IPCC	Intergovernmental Panel on Climate Change

IPPC	integrated pollution prevention and control
IPR	integrated pollution regulation
IULA	International Union of Local Authorities
IWEM	Institute of Water & Environmental Management
LBA	London Boroughs Association
LCPD	Large Combustion Plant Directive
LGIB	Local Government International Bureau
LGMB	Local Government Management Board
LIFE	EU Financial Instrument for the Environment
LLILW	long-lived intermediate level waste
LLW	low level waste
LRTAP	Long Range Transboundary Air Pollution
MAFF	Ministry of Agriculture, Fisheries and Food
MAGPs	multi-annual guidance programmes
MEPs	Members of the European Parliament
MLSs	minimum landing sizes
MMSs	minimum mesh sizes
MOX	mixed oxide fuel
MRS	monitored retrievable storage
mtC	million tonnes of carbon
NAO	National Audit Office
NCC	Nature Conservancy Council
NFFO	National Federation of Fishermen's Organisations
NFU	National Farmers' Union
NGOs	Non-Governmental Organizations
NHBC	National House Builders' Council
NHS	National Health Service
NICs	newly industrializing countries
NI	Northern Ireland
NIEG	Northern Ireland Environment Group
NIMBY	not in my back yard
NIREX	Nuclear Industry Radioactive Waste Executive
NRA	National Rivers Authority
NRPB	National Radiological Protection Board
OECD	Organization for Economic Cooperation and Development
OFFER	Office of Electricity Regulation
OFGAS	Office of Gas Services
PSBR	Public Sector Borrowing Requirement
QGA	Quasi-Governmental Agency
QMV	qualified majority voting
RCF	rock characterization facility

RECs	regional electricity companies
RPI	Retail Price Index
RSNC	Royal Society for Nature Conservation
RSPB	Royal Society for the Protection of Birds
RWMAC	Radioactive Waste Management Advisory Committee
SACs	Special Areas of Conservation
SBF	Save Britain's Fish
SCG	subcentral government
SCR	selective catalytic reduction
SEA	Single European Act
SNG	supranational government
SNH	Scottish Natural Heritage
SPRU	Science Policy Research Unit
SSSIs	Sites of Special Scientific Interest
SWPFA	Scottish White Fish Producers' Association
TACs	total allowable catches
TDs	Members of Parliament in the Irish Republic
TFOs	Third Force Organisations
THORP	Thermal Oxide Reprocessing Plant
tU	tonnes of uranium fuel
UCL	University of College London
UEA	University of East Anglia
UEG	Urban Environment Expert Group
UKCEED	United Kingdom Centre for Economic and Environmental Development
UNECE	United Nations Economic Commission for Europe
UNCED	United Nations Conference on Environment and Development
UWT	Ulster Wildlife Trust
VAT	Value Added Tax
VOCs	volatile organic compounds
W&CA	Wildlife and Countrywide Act 1981
WHO	World Health Organization
WRC	Water Research Centre
WWF	World Wide Fund for Nature

1 Introduction

Tim S. Gray

This book arises out of a Colloquium that I convened in December 1993 in the Politics Department of the University of Newcastle upon Tyne, UK.

The central theme of the Colloquium was a review of the Conservative Government's record on environmental policy since Mrs Thatcher's celebrated green speeches whilst Prime Minister in 1988. The period 1988–94 has proved to be a most extraordinary one for environmental policy in the UK, witnessing a sharp rise (1988–90) and an equally sharp fall (1990–92) in green electoral support; the White Paper, *This Common Inheritance*, in 1990 – the UK's first comprehensive policy document on the environment – was followed by the Environmental Protection Act of 1990; the Rio Earth Summit of 1992; plans proposed, shelved and revived for an Environment Agency; and the publication of the Government's long awaited policy documents on Sustainable Development (1994). The question arises as to whether or not the British Government has, during this eventful period, maintained the environmental momentum begun (or signalled) by Mrs Thatcher in 1988.

Of the many interpretations that have been advanced in answer to this question, two stand out with particular prominence. The first interpretation is that notwithstanding occasional steps backward, the period has witnessed a fundamental sea-change in UK environmental policy making, exemplified in the fact that Britain has at last formulated an overall policy on the environment. On this view, the traditional British approach to the environment – variously characterised as incrementalist, tactical, fragmented, reactive, pragmatic, minimalist, decentralised, voluntaristic, accommodatory, informal, consensual, secretive, elitist and low priority – was replaced by a new approach, variously characterised as comprehensive, strategic, integrated, proactive, principled, interventionist, centralised, legalistic, confrontational, formalised, transparent, democratised and high profile.

This change, it is argued, did not occur overnight: it was the culmination of developments during the 1980s resulting from growing pressures on the British government, especially from EC directives; from a greening public opinion; from the privatisation of the water, electricity and gas supply industries; and from an increasingly vociferous environmental lobby.

On the second interpretation, however, the only change that has occurred in UK environmental policy during the 1990s has been rhetorical. According to this view, Mrs Thatcher's speeches in 1988 were designed to capture the high ground of green credentials for the Conservative Party – to dish the Green Party – without committing the government to any fundamental re-ordering of its policy priorities. Events have subsequently borne out the far sightedness of this strategy, it is claimed, in that the electoral threat posed by the Green Party has slumped yet the government has managed to avoid both serious environmental restrictions on economic enterprise, and significantly increased public expenditure on the environment.

The bulk of the government's environmental initiatives, these critics argue, has been directed towards promoting the image, rather than the reality, of environmental concern; by, for example, banging the drum at the Rio Earth Summit, and issuing policy documents containing elegant academic appraisals of environmental problems, full of good intentions, but without any important, new, explicit commitments.

In the chapters which follow, evidence is presented supporting each of these interpretations. By studying these chapters, the reader should be able at the end of the book to assess which of these two views is more convincing as an overall verdict on the UK environmental policy during the 1990s.

Tom Burke sets the scene with a cautiously optimistic view of the UK's record of environmental policy making, arguing that the pessimists have failed to recognise the permanency of the strides made during the last 20 years to increase environmental awareness throughout the world. They have also overlooked the fact that in the 1990s, the need is for consolidation of these strides, not further innovation. Burke also challenges the view that the British people have lost their enthusiasm for environmentalism, arguing that there is an important distinction to be made between the *salience* of environmentalism, which varies very much according to immediate circumstances, and its *latency*, which is constantly, if slowly increasing. Burke believes that in general the UK's environmental record is comparable with other OECD countries, though it is often less ably defended by its advocates. However, he predicts there will be more pressure in the future on the UK government: for example, to enforce existing legislation; to introduce new regulatory legislation; to make more use of economic instruments; to formulate a more comprehensive environmental policy; to improve the public flow of environmental information; to set stricter environmental targets; and to facilitate more public participation in environmental decision-making.

Much of this pressure will come from the EU, and EU pressure is the
subject of the next chapter, contributed by Nigel Haigh and Chris Lanigan.
They show how since the mid-1980s the EC/EU has come to have a decisive impact on UK environmental policy making. After tracing recent
EC/EU developments affecting its handling of environmental issues,
Haigh and Lanigan examine various conflicts between Britain and the
EC/EU over environmental issues, predicting that such conflicts will occur
less frequently in the future. They also explain the impact of the EC/EU on
the character of British environmental policy making, concluding that
while it has forced some changes both in style (e.g., in its reliance on consensus and its decentralisation) and content, Britain has retained a considerable capacity for domestic environmental initiatives.

Neil Carter and Philip Lowe take us to the heart of the UK's environmental policy making in the 1990s by examining the thinking behind the
proposal to establish a fully integrated Environment Agency (EA) in
Britain.

Their analysis involves setting out what they see as the five central features of traditional British policy making – fragmentation, decentralisation, informality, voluntarism and closed policy community – and showing
how pressures during the late 1980s were undermining especially the
second and last of these characteristics. The proposal for an EA was
informed by a desire to integrate, centralise, formalise, professionalise and
open up the process of environmental regulation in Britain. Even more
fundamentally it reflected a strongly felt need to remove environmental
regulation from government, and thereby protect it from political pressures. A parallel debate occurred over whether to separate operational
roles from regulation, in order to prevent conflicts of interest. There was
also controversy over whether the new EA should adopt the cooperative
culture of Her Majesty's Inspectorate of Pollution (HMIP) or the interventionist culture of the National Rivers Authority (NRA). A central dilemma
in British environmental policy is thrown up by these debates: the choice
between a scientific and an administrative strategy. A further issue is
whether the EA is given extra powers (e.g., of policy participation) to
make it less of a pollution control agency and more of a genuine environmental agency.

Andrew Jordan and Timothy O'Riordan shift the debate from the institutional to the operational framework, in an extensive examination of the
most seminal principle in contemporary UK environmental policy making:
the precautionary principle. Originating in the 1970s in West Germany, the
precautionary principle's extreme ambiguity permits many different interpretations and endorses a variety of trade offs between different environ-

mental objectives. Jordan and O'Riordan provide their own interpretati
of the principle, centred on seven core elements: proaction; cost effectiv
ness; safeguarding of ecological space; intrinsic value of the environme
shifting the onus of proof to the developer; futurity of scope; and paying
ecological debts. They discuss its emergence in both Germany and the E
before reviewing its place in UK environmental practice. They show he
in some respects it sits uncomfortably with the traditionally pragmatic U
system of policy making, though in other respects it is not only consiste
with UK law and policy, but is becoming more and more applied in the U
context. Finally they provide four case studies to illustrate the applicati
of the precautionary principle in the UK.

Steven Yearley moves the discussion on to a regional basis, arguing th
in its application to Northern Ireland, recent UK environmental poli
offers a centralised perspective that is somewhat alien to the provin
Northern Ireland is a unique region of the UK, in that it is less populat
more rural, less polluted and less energy efficient with proportionally m
unemployment and industrial decline, an acute security problem,
stronger public sector, a separate legal system, a distinct political cultu
and a greater dependence on funding (including EU developme
funding). In these circumstances, environmental protection is not a ve
high priority; security and economic considerations come first. Enactme
of UK environmental legislation is delayed; implementation relies m
on consensus than enforcement; officials responsible for the environme
have lower status than their counterparts in transport or planning; there i
lack of any effective voluntary sector not least because environmen
non-governmental organisations (NGOs) are part funded by t
Department of the Environment in Northen Ireland (DoE NI); and t
people of NI are relatively less concerned about environmental issues.
the 1990s some new developments served to alter this situation: f
example, a new Environment Service was formed within the DoE NI an
new advisory committee – the Council for Nature Conservation and t
Countryside, or CNCC – was established. But doubts exist as to the effe
tiveness of each of these bodies in moving the environment up the politi
agenda in Northern Ireland. At best, environmental initiatives in Northe
Ireland seem destined to continue to serve economic interests rather th
the other way round. Moreover, it seems likely that the environmen
policy agenda will continue to be set by Whitehall not Belfast.

Stephen Ward continues the subnational focus, by a study of the loc
level of UK environmental policy. The central theme of Ward's paper
the increasing linkage between the local and the European level of U
environmentalism. This linkage has been encouraged by both sides – loc

uthorities and the European Commission – mainly because of their espective frustrations with the national level of UK environmental policy naking. Local authorities feel that their views receive more attention in 3russels than London, while the Commission obtains helpful advice from he grass roots implementers of EC directives. Moreover, by this linkage ooth bodies gain an added legitimation for their environmental policies, hereby reducing their respective democratic deficits. However, the inkage is weakened both by the negative attitude of the UK central gov-:rnment, and by the fragmented nature of local governments' contacts with European bodies. Nevertheless some local authorities are forging ahead with their European contacts, especially those authorities in which .enior officers take the lead, which have a European strategy, which seek EU environmental funding, which are active environmentally, and which are large in size.

Turning now to particular policy areas, David Maddison and David Pearce discuss UK policy towards global warming. UK ratification of the Rio Earth Summit's 1992 Climate Change Treaty committed the country to stabilise greenhouse gas emissions to 1990 levels by the year 2000. Maddison and Pearce point out that although the government had previously undertaken to use economic instruments to attain this target, the measures it has recently taken (imposing value added tax – VAT – on domestic energy and increasing road fuel duty) are unlikely to achieve all the savings required. Eventually the government may have to overcome its objection to the introduction of a European carbon energy tax (based on its fear of loss of fiscal sovereignty if tax decisions are made in Brussels). If and when it does accept a carbon tax, it may find that such a measure not only delivers more savings of greenhouse gas emissions, but also has a positive effect on both gross national product (GNP) and employment. Maddison and Pearce conclude that 'Carbon taxes are down, but not out.'

Gerald Manners contributes a more discursive view of the UK's energy conservation policy, setting recent developments in the context of the last 20 years of shifts in decision making. Manners describes the potential benefits to the UK of a vigorous policy of energy conservation, and analyses the dilemma of a non-interventionist government in taking the initiative in promoting measures for energy saving. By 1993 these misgivings were put to one side, and the government accepted its central role in energy conservation, though it still faced much political opposition to a key element in this stance – the rise of energy prices – not least because of the expectations of low prices following privatisation of the gas and electricity industries. Manners explains the activities of the Energy Saving Trust, but argues that more measures are needed (both regulations and

market instruments) if energy efficient investment is to be significantl
increased. Hence, although government energy policy in 1994 was mor
coherent than in the past, there are still crucial gaps between aspiratio
and performance, and the UK energy ratio remains disappointingly highe
than that of many other OECD countries. Paradoxically, target setting fc
emission reductions may not optimalise energy saving, and the govern
ment's public spending parsimony may undermine the effectiveness o
some of its policy measures.

This theme of energy conservation is applied by Elizabeth Shove to th
construction industry. She warns against the oversimplified assumptio
that only by legislation, and not by persuasion or by financial incentives
can operators be brought to reduce their energy consumption. Regulatio
is a complex process of relationships, and no outcome can be guaranteed
This is nowhere more true than in the construction industry, which is a
major energy player since half of all carbon dioxide emissions arise out o
energy consumption in buildings. Shove points out that health and safety
regulations are easier to enact and enforce than environmental regulations
This is partly because the capacity of the construction industry to comply
with radical new environmental requirements is extremely limited, at leas
in the short run, for both technical and economic reasons. It is also
because the rationale for these environmental requirements is highly dis-
puted:for example, builders reject cavity insulation because 'houses
should be able to breathe'. Hence it is hardly surprising that current energy
saving performance in the British construction industry is thirty years
behind that of Scandinavia.

A closely connected area of UK environmental policy, transport, is
examined by Kenneth Button. He points out that transport has always
raised environmental problems, but that more recently with the sharp rise
in road traffic, these problems have become both more acute and more
widespread, reaching beyond national borders. After analysing the environmental damage caused by transport, Button shows how initiatives taken
to deal with this damage – even when successful – may well entail negative trade offs (i.e., a worsening of the environment in other respects). The
UK government's response has traditionally been focused on planning and
technical controls with little reliance on fiscal measures. However, in the
1990s this response has changed significantly, largely because of political
resistance to further sharp rises in road traffic. Basically the new response
is to restrict demand to the existing supply, rather than, as before, to
expand supply to meet an ever rising demand. In doing so, the UK is
acting in concert with the EU, and in line with the Rio Earth Summit
Global Climate Convention. However, some critics view the UK's new

iscal policy measures (such as increased fuel duty) as driven less by environmental considerations than by the need to raise revenue and thereby reduce the public sector borrowing requirement (PSBR).

The 1990s saw a change in the opposite direction in UK policy towards acid rain. Jim Skea points out how the UK was dubbed the 'dirty man of Europe' because of its failure to take effective action to reduce its acid emissions. He claims that this issue was so controversial that it became the UK's defining environmental issue during the 1980s. In 1988, however, the EC's Large Combustion Plant Directive (LCPD) ended the political debate (i.e., whether sulphur dioxide emissions should be cut, and if so by how much) but it did not end the administrative skirmishing over the means of achieving the agreed cuts. After reviewing the respective positions of key actors on the issue, Skea analyses the evolution of UK policy in three periods: pre-1988; 1988–1993; and 1994 onwards. Until the late 1980s the UK government chose to resist dealing with the problem of acid rain. After 1988 it was required to take action, but it still insisted on the need for strong scientific evidence and on the importance of taking into account the economic cost of remedial action. During the 1990s, however, the issue of acid rain has faded in significance in UK environmental policy, partly because it has been eclipsed by the issue of climate change, and partly because the LCPD appeared to have dealt satisfactorily with the problem. Ironically, however, the UK is likely to achieve its sulphur dioxide emission reduction targets not because of environmental considerations, but because of its financial policy towards the energy industry. In the future the UK is likely to face increasing pressure from the EU to increase its sulphur dioxide emission reduction targets; without such pressure, UK acid rain policy is unlikely to recover its earlier salience.

Andrew Blowers's chapter brings us up to date on another highly contentious issue of the 1980s and 1990s: that of nuclear waste disposal. Blowers shows how the British Government's policy on nuclear waste disposal is currently in disarray, as the waste has accumulated but options for its disposal have run out, one by one, in the face of intense public opposition. He analyses in turn the technical, economic, environmental and social dimensions of the conflict, before going on to examine in detail the phases of political decision making over the issue. During these phases the traditionally secretive culture of nuclear decision making has been challenged by increasing demands made by anti-nuclear groups for greater participation, and there has been a growing internationalisation of the issue of nuclear waste disposal. These developments reflect the fact that the debate has moved on from purely technical and scientific issues in the 1960s to highly politicised issues from the 1970s onwards. As a result of

NGO and international opposition, the option of sea dumping had to be abandoned in the early 1980s, and because of local community opposition the Nuclear Industry Radioactive Waste Executive (NIREX) has to date failed to secure any land based site outside Sellafield itself. In the 1990s controversy concentrated on the Thermal Oxide Reprocessing Plant (THORP). By the time THORP had reached the commissioning stage although much of its technical and strategic rationale had been eroded, the government decided on economic and social grounds to give it the go-ahead. This decision does not however end the controversy; serious political difficulties remain with the problem of foreign wastes and the future prospect of the NIREX repository. Deep uncertainty continues to lie at the heart of UK radioactive waste policy.

Moving from issues of pollution to issues of conservation, Stephen C Young assesses the UK's wildlife policy since 1988. The 1981 Wildlife and Countryside Act is the foundation stone of current British biodiversity policy, affording protection both to specific geographical areas – Special Sites of Scientific Interest (SSSI) – and to 400 endangered species of flora and fauna. However, the Thatcher Government was more concerned with deregulating the economy than conserving wildlife, and as a result, many important wildlife sites were damaged. The Nature Conservancy Council (NCC) was powerless to prevent this process. However, the 1986 Agriculture Act to some extent redressed the balance by imposing regulatory controls on farming practices, and ending the post war period of agricultural expansion. Under the Environmentally Sensitive Areas (ESAs) scheme, farmers were paid to manage the countryside in an environmentally sensitive manner, not to maximise production. In 1985 a similar responsibility was placed on the Forestry Commission. Moreover, significant wildlife sites in urban areas were identified for protection for recreational purposes. However, there was conflict within central government: some departments (such as Transport – DoT, Trade and Industry – DTI, and even DoE) were undermining wildlife protection, whereas others (notably the Ministry for Agriculture, Fisheries and Food – MAFF) were moving towards protection. Moreover the separation of the NCC into separate organisations for England, Scotland and Wales weakened its national impact. On the other hand, the splitting-up of water regulation from water supply in 1989, and the parallel splitting-up of forestry regulation from forestry management in 1991, were promising moves environmentally. However, since 1988 there has been little major policy shift in UK wildlife policy. Some initiatives, such as the ESAs, were extended; there has been rigorous enforcement by the NRA of pollution controls on farming practices; and more species have been added to the protected lists. However, many more protests about adverse impacts on wildlife of economic development have been ignored than accepted, and the general direction of government wildlife policy has remained centred on the voluntary principle rather than on regulation, as in other areas of government policy. As a result, the negative trends have been exacerbated, only partly compensated by the positive trends, and 'an air of uncertainty hangs over the future of conservation policy'.

Fish are wildlife, though subject to systematic harvesting for human consumption. The conservation of fish stocks around the UK for purposes of commercial exploitation is the subject of the chapter by Anthony Stenson and Tim S. Gray. Here the focus is on the Common Fisheries Policy (CFP) of the EU, to which the UK has been subject since 1972. Stenson and Gray argue that the CFP has been a 'tragedy', not in the full blown Hardinian sense of the complete collapse of the commons, but in its failure to maintain a system of sustainable development. Fish stocks are overexploited, and there is too much capacity in the fishing fleet. During the 1990s the problems have become more acute, and the government's response – the days-at-sea scheme under the Sea Fish (Conservation) Act 1992 – has served only to inflame an already explosive situation. Stenson and Gray explore the historical and political roots of the crisis, and track the oppositional strategies of the fishermen's organisations, including a successful legal action which has produced a (temporary) suspension of the days-at-sea scheme. They show how, without a common financial interest such as that which lies at the basis of the Common Agricultural Policy (CAP), the EU's CFP is almost bound to fail in its central objective of conserving fish stocks to sustain a large and healthy fishing industry.

Michael Redclift concludes the book by setting UK environmental policy during the 1990s in the context of global environmental problems. He notes that the British government has frequently voiced concern for the global environment, but this concern has not yet led to action to tackle the fundamental root of global damage: the structure of international economic relationships. Redclift points out that the British Government is not alone in ignoring the root of the problem: at Rio the central issues of poverty, trade, the debt crisis, the water crisis, desertification and population were neglected by most northern governments. As a result the outcome of Rio was a set of commitments (e.g., on global warming and biodiversity) which were imprecise and failed to secure the necessary financial basis of transfer of funds from the North to the South. Moreover, the UK's domestic policies – especially on emission targets, energy conservation and biodiversity – have lacked the sense of urgency that Rio demanded. In Redclift's view, the documents published in January 1994

setting out the UK strategy on sustainable development fall well short of
an adequate response to Rio. He concludes by suggesting ways of improving that response, including greater commitment to the EU's Fifth
Environmental Action programme (by, for example, endorsing an EU
carbon tax and increasing 3rd World funding); in short, a greater resolve
to enter into the spirit of the Rio international agreements which we have
after all, recently ratified.

Although it is difficult to summarise such a diverse set of evaluations of
UK environmental policy since 1988, three conclusions do seem to
suggest themselves.

First, while some progress has undoubtedly been made, there has not
been a fundamental paradigm shift in environmental policy as a result of
Mrs Thatcher's 'green' speeches.

Second, many environmental improvements have been due to measures
imposed by the EU and reluctantly implemented in the UK, rather than
independently chosen by the UK Government.

Third, it is not yet clear whether the government's performance during
the remainder of the 1990s will match its present commitments and future
aspirations.

In short, neither of the two interpretations advanced earlier seems convincing. We have not witnessed a fundamental sea-change in UK environmental policy, but the government has exhibited more than merely
rhetorical concern for the environment. There is a fragile momentum
building up towards the development of a coherent environmental policy:
only time and circumstance will tell us whether this momentum is to be
sustained or extinguished by the UK Government in the next millennium.

2 View from the Inside: UK Environmental Policy Seen from a Practitioner's Perspective

Tom Burke

My perspective is that of a practitioner, not a student, of UK environmental policy. I have spent the last 20 years trying to *do things* about environmental problems rather than to simply *study them*. I have been fortunate in having had a chance to do that both from outside Government, as Director of Green Alliance (and formerly, of Friends of the Earth), and for the past few years from inside Government, as Special Adviser to the last three Secretaries of State for the Environment. This is a transition of roles which is sadly unusual in the British political system. As a result, I do not pretend to give here an objective or analytical view of UK environmental policy. Indeed, I am rather sceptical about attempts to subject the processes of environmental politics and policy to objective analysis in terms of precise models (see below for my scepticism about economic models developed to substantiate the case for market instruments as a substitute for command and control methods of environmental regulation). I am more comfortable with the considerable untidiness of actual events.

I am also unsympathetic to the tendency on the part of some environmentalists to take a morbid fascination in our inability to prescribe effective remedies for the ills they perceive around them. In fact, I resent the ghetto mentality of environmentalists which produces this sort of morbidity. I take a more optimistic – in many ways more realistic – view of our recent handling of environmental problems. The endless picking away at sores, to which many in the environment business are prone, makes us constantly feel we are failing. We have only to listen to some environmentalists to think that we have accomplished very little in the last 20 years. This is a distorted perspective, and both underestimates the problem of where we have come from, and undermines our capacity to go forward.

The UN Conference on Environment and Development in Rio in 1992 is an example of a recent event which brings out the worst in those who tend towards methodical tidy-mindedness and pessimism, and thus create

11

a false picture of where we really are. Rio is a particularly appropriate illustration, since the thrust of this edited book is an appraisal of UK environmental policy in the wake of the euphoria of the Earth Summit. According to some of the noisiest critics, Rio was a failure, in that its momentum has not been carried through into real policy changes. I want to issue a salutary warning against this way of thinking, by recalling that exactly the same views were expressed soon after the UN Conference on the Human Environment in Stockholm in 1972, and yet enormous strides in environmentalism have taken place since that conference 23 years ago.

In order to grasp the momentous nature of this change, let us compare the extent of environmentalism in 1972 with that in 1992. On the eve of the Stockholm Conference there was relatively low public awareness about the environment; there had been little or no real analysis done of environmental problems, and what analysis there was was fragmentary and certainly not *policy* analysis but only *problem* analysis; there was practically no international legislation, and in most countries little national legislation; there were few institutions with any strength internationally or nationally for dealing with environmental policy; and there was no such thing as an environmental press corps. The politics of the environment, such as it was, was confined to a small group of scientists, journalists and environmentalists talking largely to each other.

By contrast, by the time of the eve of the Rio Conference there was a huge public awareness of environmental issues; a deep analysis of the problems and their connections to other issues and a coherent structure of ideas by which to interpret environmental events; an enormous amount of international and national legislation (much of it admittedly observed mainly in the breach); many environmental institutions, both international (especially the European Union) and national; and an extensive environmental media corps. In short, the environment had by 1992 become a part – albeit a junior part – of mainstream politics all over the world, instead of being the more or less exclusive preserve of a small band of enthusiasts and missionaries.

In my view this change was to a significant extent due to the influence of the Stockholm Conference. Its influence was not immediate; it did not happen the day after the conference ended. But the conference's message gradually filtered into global consciousness through the infinitely complex channels of bureaucratic and other institutional communication. Thus in attempting to evaluate the Earth Summit at Rio, we should recognise that it is probably much too early to get any sense of what Rio has really accomplished in terms of policy transformations. The best (and most) is yet to come.

This means that we are now entering a consolidating stage of environmentalism. Some people's frustration with the aftermath of Rio is due to their irritation that we are no longer innovating. But we are now in the implementation stage, not in the innovation stage. Moreover, there was a real danger that we were using innovation as a substitute for action (i.e., implementation) and it is right that there should now be a pause. In this sense Rio was the end of something, not the beginning of something. It was the end of an intellectual process that started in the OECD in the late 1970s with the development of ideas about anticipatory and preventative policies, and ran through the World Conservation Strategy and the Global 2000 report, culminating in the Brundtland Report, which was the inspiration of the Rio Earth Summit. The real accomplishment of Brundtland was not so much to produce a synthesis of these earlier ideas, but to translate them into a political process with a momentum that drove along the Earth Summit.

Rio, then, is to be seen as drawing on the intellectual capital of past explorations, not as creating new environmental ideas. The products of Rio – the Conventions on Climate Change and Biodiversity, the Declaration of Principles, and Agenda 21 – were transformations of that intellectual capital into programmatic action. Thus our immediate task is very much what it was after Stockholm, to concentrate on implementation, which entails a hard, frustrating slog before we begin again to think in creative terms; to accumulate a capital of practical experience with which to refresh our ideas.

There is another respect in which gloom is misplaced. Many critics argue that with the economic recession, public enthusiasm for environmentalism has waned, and with it the government's commitment to green policies. But the fact is that whatever the impression given by the newspapers, throughout OECD countries public concern about the environment remains extraordinarily high: indeed at historically high levels, despite the recession. For example, European polls indicated that the average level of concern about the environment went up 11 per cent between 1988 and 1992. Typically 80 per cent of the population in developed countries express a high level of concern about their environment. This suggests to me that public expectations of effective governmental action on the environment are probably greater than they have ever been. There is no sign of any shifting back to lower levels of concern. I suspect that this is mainly because this public concern is driven by practical experience rather than by the power of ideas. In other words, it is not a conceptual debate that is going on in the public mind; rather it is the personal (and often intimate) experience that so many people have with an environment that is deteriorating, becoming noisier, dirtier and more dangerous to health. Everyone is influenced by scares about, for example, air pollution and asthmatic risks to children, sewage in bathing water, and radioactive leakages. Television heightens our awareness of global problems. We have all sat in our living rooms watching the rainforest burn live on our televisions, the oil floating down the Gulf, the hole in the ozone layer (blue with wiggly edges) and the bodies around Bhopal. These surrogate experiences of a planet in peril reinforce and legitimise our sense of a personal environment under threat.

It is true that there are periodic shifts up and down in the relative weight of popular concern over the environment as expressed in snapshot opinion polls. But we must distinguish between *salience* (the immediate level of popular concern, which necessarily varies from time to time as different issues hit the headlines) and *latency* (the underlying level of popular concern, which is constantly, if slowly, increasing). In other words there is a high latency of public concern about the environment, but its salience moves up and down, as salience always does, triggered by events of the moment. This means that although the immediate pressure on the government for environmental improvement may vary, there is a general feeling among the public that the government's action is too little and too late. Polling evidence indicates that the public sees a big performance gap between what they are encouraged to expect and what they actually get. I think that in the near future – say, over the next five years – a great deal of what will shape governmental environmental policy will be the pressure to close this perceived performance gap.

This pressure will probably take the following forms. *First* there will be intensified pressure to enforce existing legislation more vigorously. This pressure is already having some effect in encouraging HMIP and the NRA to prosecute more offenders. Such a response is attractive to governments since it costs them little to beef up the regulatory machinery; indeed, the increased income from fines may help to defray the cost of the regulation. I can even see the point being reached when very harsh action will be taken against persistent pollutant offenders, including shutting down their operations for a time. Nothing reassures the public more than seeing a few bodies hung from the battlements and the more prominent and better known the bodies, the more salutary the effect.

Second, there will be increased pressure to introduce new legislation to plug the gaps that remain in UK environmental regulation. This pressure will come mainly from the EU, though not, perhaps, as vigorously as it has in the past, at least for a few years. I suspect that the pace of EU environmental legislation will slow somewhat in the aftermath of Maastricht, as the subsidiarity principle begins to bite, but the slowdown will only be temporary, as the Clinton/Gore administration begins to speed up its environmental legislation, and there develops some sort of bidding pressure between the EU and the USA for leadership in environmental matters.

Third, there will be increased pressure to introduce market instruments as an alternative to regulation. However, this pressure is unlikely to result in a wholesale substitution of economic instruments for regulation: at best we will end up with a mixture of both. Moreover, we should not expect that the revenue from such economic instruments will be accumulated in the DoE for expenditure on environmental objectives. I have to say that I am a deep sceptic about the value of economic instruments unless they lead to the creation of real markets that generate real prices. By and large we have very little empirical evidence that they work; indeed, it is difficult to know how we can tell whether or not they work. For instance, how can we determine whether the recent imposition of VAT on fuel will actually contribute to our reaching our climate change target? In the much vaunted case of lead-free petrol, what was important in persuading people to switch to lead-free petrol was not the lower price but, rather the availability of lead-free petrol on the garage forecourt. In general, conformity with economically rational behaviour tends to follow the degree of corporate, as distinct from consumer, choice. Large companies tend to conform more to theoretical models of economically rational behaviour than do small companies who, in turn, are by and large more rational than families or individuals. In my view, all that you can confidently say about economic instruments is that they will raise lots and lots of revenue, not that they will achieve environmental objectives. Moreover, I suspect that environmentalists will be too weakly placed to ensure that economic instruments are designed to optimise environmental, as opposed to revenue, outcomes.

However, there is one kind of pressure from economists that will be impossible to resist: pressure on environmentalists to learn a lot more about the macrodynamics of the economy. If the microeconomics of market instruments is pretty marginal to influencing environmental outcomes, macroeconomic questions, such as the right level of inflation or the right price of money or the right volume of trade or the right rate of growth, for sustainable development, are vital questions to environmental policy making, and environmentalists will be under increasing pressure to answer them.

Fourth, since Rio there has been increased pressure on the government to formulate a comprehensive environmental policy for sustainable development. This pressure is likely to intensify, and it will have two effects: (a) it will further establish the environment as a legitimate area of government policy making; (b) it will enhance the status and power of the DoE, consolidating its control over the machinery for environmental regulation, and giving it a charter to increase its influence over the policies of other departments of state, especially that of the DoT. This is likely to mark the end of the situation in which the DoE has been relatively weak, fragmented and underresourced in both personnel and budgetary terms, and to make it more attractive as a career choice to the more able civil servants and politicians.

Fifth, there will be more pressure on the government to set targets for industrial, manufacturing and construction operations, along the American lines, so that improvements in performance can be tracked empirically. More regular and systematic reports on these achievements will also be demanded.

Sixth, there will be increased pressure to improve the flow of environmental information into the public domain. This pressure for greater transparency will come partly from central government itself (part of the current style of open government), and partly from NGOs, but most of all from the financial institutions in the City. Increasingly investors, insurers and lenders are coming to recognise that those who manage the environmental risk of companies are managing their environmental exposure to a wide array of potential liabilities.

Seventh, we can expect increased pressure for more effective public participation in decision making. This pressure will come from both established NGOs and *ad hoc* local voluntary groups, motivated partly by NIMBY (not in my back yard) considerations and partly by the desire to become involved more in decisions which they see as directly affecting their lives.

How well equipped is the UK to deal with these pressures? In terms of the *machinery* of government, it is probably better equipped than most other countries, in that the elements of a comprehensive approach to environmental policy are now in place, with the sustainable development strategy together with an environmental regulatory system charged with the responsibility of implementing integrated pollution control, and a legal regime for public information dissemination. All the elements are there, though whether they are yoked into effective action depends on political will. In terms of its past record of *policy making*, the UK's environmental performance is also better than most, although as in many other countries it is patchy. But in terms of the *politics* of environmentalism the UK does extremely poorly. We are almost as bad as the Americans under President Bush who, despite having by far the best national strategy on biodiversity protection, managed to provoke outrage by not signing the biodiversity treaty at Rio. In the UK we are sometimes similarly inept in the way in which we present ourselves environmentally, especially (but not exclusively) to the outside world.

Note

This chapter reflects the opinions of the author, and should not be interpreted as expressing the views of the UK DoE.

3 Impact of the European Union on UK Environmental Policy Making

Nigel Haigh and Chris Lanigan

INTRODUCTION

In early 1982 the Campaign for Lead Free Air (CLEAR) was launched with the objective of removing lead from petrol. The campaign was well funded; it was headed by an experienced campaigner; and it was backed by an impressive array of political and scientific figures and environmental groups. It was also successful. Within a little over a year the British Government had changed its policy to one of support for lead-free petrol and two years later an EC Directive was adopted requiring the supply of lead-free petrol to be mandatory throughout the EC from October 1989.[1]

A surprising feature of the first number of the glossy newspaper issued by CLEAR, is that it nowhere mentions the existence of EC legislation that then prevented member states from insisting on lead free petrol. In launching his campaign, Mr Des Wilson did not know of the existence of the EC Directive. Neither did he then realise that if his campaign was to be successful he would have to persuade the EC Commission to propose an amendment to the Directive and obtain the agreement of all the other member states. To give him his due, he very quickly learnt and with his usual energy proceeded to 'Europeanise' the campaign.[2]

How is it that some nine years after the EC had decided to embark on an environmental policy, by which time a substantial array of important EC environmental legislation had already been adopted, it was possible for an experienced environmentalist, and one who was a leading figure in a political party that had always insisted on its commitment to the European idea, to plan a campaign in purely national terms? One can only conclude that the power of the EC to determine national environment policy was not then recognised even by the politically active, let alone by the wider public.

A decade later the position is quite different. National environmental bodies rely on EC legislation in their daily work as a matter of course, as do the regulatory bodies. Industry has organised itself to get its voice heard in the making of EU legislation. Government officials and even Parliamentary Counsel have been known to travel to Brussels to check clauses in Parliamentary Bills with Commission officials. The public, according to social surveys, is aware of EC environmental policy and is supportive of it. By 1990 it was possible for Mr Derek Osborn, Director-General of Environmental Protection at the Department of the Environment to say:

> In a relatively brief period of time the nature of the Community has evolved and expanded, so that its many strands are now woven into most areas of public policy. This is certainly true of environmental policy, where in the last 20 years the European Community has rapidly become a major factor, or even the dominant one.[3]

This chapter attempts to explain this change and is in three main parts. First, it will set out some developments in the way in which EC/EU environmental policy is handled. Second, it discusses recent conflicts between the EC/EU and Britain over environmental policy. Third, it analyses the effects of the EC/EU on the style and nature of British environmental policy making.

DEVELOPMENTS IN THE HANDLING OF EC POLICY

In the first ten years of EC environmental policy (1973–1983) EC institutions concentrated on the adoption of a quantity of legislation and very little effort was made to consider the impact it was having. The steady growth in quantity is shown in the Figure 3.1 (taken from Haigh, n.d.). Mere numbers say nothing about the importance of the items, and while some are of a narrow technical character (often consisting of minor amendments) there have been important items of legislation adopted in most years.

The first book to be published on EC environmental policy had very little to say on its implementation (Johnson, 1979). Indeed, the first edition of this book said nothing, while the second edition of 1983 devoted only two pages out of 244 to the subject. The first sustained discussion of the implementation of EC legislation in one country and the impact it had had did not appear until 1984 (Haigh, 1984a).

In 1983 a dramatic event drew public attention to the importance of implementing EC legislation. Some drums of hazardous waste thought to contain dioxin and originating from Seveso near Milan (where a major

Figure 3.1   Number of items of EC environmental legislation adopted each year excluding those relating to radioactivity

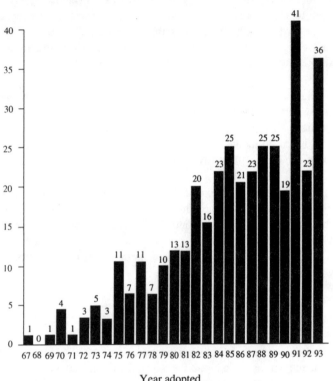

Year adopted

accident had happened in 1976) went missing, and were later found in Northern France. As a result, the European Parliament appointed a Committee of Inquiry to investigate the implementation of an EC Directive of 1978 on toxic waste. This committee came out with forceful and witty conclusions ('In the light of replies from the Commission, which were inadequate, and from the Council, which were dilatory …').[4] As a result the European Parliament adopted a Resolution which censured the Commission for having failed to perform fully and properly its role of guardian of the Treaties and for failing to take the necessary measures *vis-à-vis* the member states with regard to the implementation of the Directive.[5]

From then on, implementation became a much greater concern of the EC Commission, and the staff of the legal unit of the Directorate General responsible for the environment (DG XI) was expanded. The head of the unit who was then appointed, Dr Ludwig Kramer, has now become one of the best known of the officials of DG XI, and his vigour in recommending that member states be brought before the European Court has focused far greater attention on its potential effects.

The next major development was the drafting of the Single European Act, which came into force in 1987. This not only legitimised EC environmental policy by providing a clear legal base but also introduced at least three other important elements.

First, the hand of DG XI was substantially reinforced by the statement in the Treaty that 'environmental protection requirements shall be a component of the Community's other policies'. Second, legislation relating to the single market could be adopted by qualified majority voting (QMV) in the Council with the result that no one country (indeed no two countries) could block legislation against the will of the others. This significantly alters the negotiating style of member states in the Council since no one country wants to find itself isolated. Third, the powers of the European Parliament and the Commission were greatly strengthened by the introduction of the so-called 'cooperation procedure' which gives the Parliament the opportunity of a second reading in certain circumstances, and enables it, if supported by the Commission, effectively to force the hand of the Council. The Parliament did this with standards for vehicle emissions, as a result of which catalytic converters on cars have effectively become obligatory.

Then came the Treaty of Maastricht which was intended to lay the foundations of political and monetary union but which, among its other innovations, extends the range of environmental matters which are to be decided by qualified majority in the Council. It also further strengthens the powers of the Parliament with the introduction of the co-decision procedure (Wilkinson, 1992).

RECENT CONFLICTS BETWEEN EC AND UK OVER ENVIRONMENTAL POLICY

Historical Context

EC environmental policy dates from 1973, when the first Action Programme was agreed. However, the environment was only beginning to become important during the 1970s and DG XI was politically weak and underresourced. Early environmental Directives such as those on bathing water (1976), birds (1979), and drinking water (1980) owed their existence to their lack of conflict with the interests of the established and powerful Directorates. British approval was given with little hesitation: the protection of birds was a long-standing policy in Britain, and the idea that British water might not be clean enough to pass tests which would also have to be met by continentals with supposedly dirtier water probably did not occur to the British Government. In any case, the five-year period allowed for implementation allowed time for any necessary adjustment, and there was always the probability that delays and derogations (i.e., exemptions) would be granted in difficult cases.

Developments during the 1980s drastically changed this picture. In addition to the strengthening of the legal basis for Community action on the environment as noted above, the internal standing of DG XI within the Commission was increasing. This was partially due to the fairly impressive list of Directives that had already been agreed which lessened the previous tendency within the Commission as a whole to view DG XI as weak, unimportant and peripheral. It had gained the respect and experience necessary to make tough decisions that affected the interests of other Directorates, and indeed those of member states' governments.

This heightened respect and experience became salient as mounting public concern over the environment (due to high profile disasters such as Chernobyl, dire warnings about the future effects of global warming and the decline in the ozone layer) encouraged the belief that the EC had a major role to play in environmental protection. These concerns were linked with more local and narrower issues such as the effects of leaded petrol, and growing pressure to develop rural areas. Green parties were making headway in the more developed parts of the EC, and environmental pressure groups were growing in both numbers and impact. This increased the pressure on governments to agree to proposed European level legislation, and encouraged the Commission as a whole to emphasise the environment in its programme, and to support the actions of DG XI's officials.

The arrival in 1989 of the flamboyant and outspoken Italian socialist Carlo Ripa di Meana in the post of EC Environment Commissioner, coupled with the growing influence of 'Euro-sceptics' in the British Conservative Party and DG XI's new found emphasis on implementing existing environmental Directives, created a potentially explosive mixture which in many respects has coloured the current relationship between Britain and the Commission over EC/EU environmental policy, and, to some extent, the very content of that policy.

Britain and Compliance with European Directives

The early environmental Directives were negotiated by the British Government during a time of International Monetary Fund (IMF)-influenced public spending cuts, coinciding with great difficulties for the government in forcing major, time-consuming bills through Parliament. Even the incoming Conservative Government in 1979 with its safe majority faced pressures on parliamentary time. This encouraged the use of departmental circulars and Statutory Instruments (under the European Communities Act 1972) rather than new bills to implement Directives. The Commission eventually forced the government to discontinue the use of mere departmental circulars, but as Statutory Instruments are still considerably less time-consuming than introducing new Acts of Parliament they have become the norm for implementing the requirements of Directives. Directives were seen by the British Government as flexible instruments, the implementation of which could take considerations of finance, time and vested interests into account. It was the decision of the Commission to challenge aspects of this approach which led to the clashes between itself and the British Government.

The traditional approach can be seen if we examine the Environmental Impact Assessment Directive. Formally proposed by the Commission in 1980, it only became EC law in 1985 after the British Government and the Danish Parliament had scaled down its provisions and exempted projects resulting from acts of national Parliaments. No new British Act of Parliament was passed. Instead, the government issued a number of Statutory Instruments and published advisory Department of the Environment circulars, designed to influence the planning permission system. Many of these, especially those for Northern Ireland, were not made until the early 1990s. Initial draft DoE circulars took the view that most categories falling in the large 'grey area' of annex 2 of the Directive did not require impact assessments as far as Britain was concerned, and the government refused to accept the Commission's interpretation that the Directive should be applied to projects which had entered the initial stages of the planning process prior to the Directive coming into force.

Rather than pursuing a conciliatory approach, which would have involved gently encouraging the British Government to accelerate the progress already being made in achieving Directive targets, and waiting for test cases to confirm the legal meaning of ambiguous parts of Directive wording, DG XI chose to put heavy pressure on the British Government to conform to DG XI's interpretation of the Directives, and to do so by the previously agreed compliance dates. Action would be taken in the Court of Justice if Britain's responses were deemed unsatisfactory.

The Commission forced the British Government to amend draft DoE circulars to planners and as a result many more development projects than the government had wished became the subject of impact assessments. The Commission took up the complaints of pressure groups and issued 'Article 169' letters calling for the British Government to respond to the allegations that it was breaking the provisions of the Directive in several cases, including the controversial Twyford Down and Oxleas Wood road schemes. This same period (1989–92) also saw the UK threatened with Court action over polluted bathing waters, the low number of air pollution monitoring stations and the shooting of 'pest' birds. This extraordinary series of disagreements between the Commission and the UK Government requires explanation. Perhaps surprisingly, this explanation owes much to psychology and personality in addition to politics and the actual environment.

In the mid-1980s both government and many ordinary people in Britain generally held to the belief that Britain was one of the most ecologically conscious countries in Europe. There was also a belief that Britain was more scrupulous than many other Community states in carrying out international obligations agreed to. This self-image made the actions of DG XI over standards and enforcement of Directives hard to digest. It was easy to rationalise the situation to conform with the self-image by believing that it was Commission interference rather than British standards which was the problem.

DG XI's actions owed much to similarly strong beliefs, in particular those of Ludwig Kramer, the head of its enforcement section, who has publicly expressed his views on the importance of law as an active instrument in the battle to safeguard the environment. Briefly stated, he considers the laws as they are on the environment to be insufficient for their task as it is, and given this he is unwilling to allow political considerations and 'understandings' to reduce the scope for improvement already present. For him, Directives should be interpreted and then enforced in ways which maximise benefits for the environment, and the role of the DG XI is to act for the mass public and indeed the general interest of future generations who lack the resources, information or representation to check and balance the machinations of governments and vested interests.[6]

Therefore, when the British Government came to realise the huge costs and political implications of what they had agreed, they found themselves dealing with officials unwilling to acquiesce in a flexible interpretation of Directive standards, compliance dates or monitoring requirements.

Decisions to initiate infringement proceedings before the Court of Justice are taken by the Commission and not by officials; but the environment Commissioner was now Ripa di Meana, and he was willing to take a strong line for two reasons. Firstly, he too was convinced of the importance to the environment of taking a maximalist view of EC Directives and strictly enforcing this interpretation. This view would not have been shared by several of the other Commissioners if they had been given the post (it is, for example, hard to imagine Jacques Delors or Leon Brittan giving the same enthusiastic support to the cause of environmentalism).

Secondly, it seems likely that Ripa di Meana came to favour high profile action as a means of compensating for his lack of status within the Commission as a whole. Gaining media coverage, and through it a reputation as a heavyweight politician standing up both for environmental and Community interests, clearly had both political and personal attractions. Having previously had a period of residence in London, Ripa di Meana was well aware of the British self-image of itself with relation to the EC and the environment. Attacking British 'failures' on the implementation of environmental Directives was inevitably going to cause more of a stir than similar action would have caused in other EC states,[7] but he saw this as an opportunity to be taken, rather than a problem to avoid.

Ripa di Meana was prepared to go to unusual lengths in achieving these aims. Perhaps the most notable case occurred as a result of the disagreement between the Commission and the British Government over whether the contentious Twyford Down road scheme met (or even needed to meet) the requirements of the Impact Assessment Directive. Ripa di Meana persuaded the College of Commissioners to allow him to attach a letter requesting a halt to construction work to the normal, standard 'Article 169' notice of investigation. This agreement was given on the understanding that the affair would be conducted with maximum discretion and without publicity. Ripa di Meana proceeded to brief Members of the European Parliament (MEPs) and his staff sent a copy of the letter to the Commission's public relations department for circulation to the media![8]

There were also more straightforward political factors which fuelled conflict between the protagonists. The privatisation of the English and Welsh water companies was a particular problem, partly because the Commission was unwilling to treat private companies as 'competent authorities' to monitor and implement standards (monitoring and enforcement were eventually entrusted to the NRA) and partly because both bathing and drinking water Directives required large capital investment if standards were to be met quickly, and this threatened to discourage potential investors.

In addition, it appeared for a time that environmental issues might have a significant influence on voting patterns in Britain. In retrospect, the 15%

vote which the Green Party attained in the 1989 European elections can be seen to reflect the temporary collapse of a viable 'third party' alternative, and the perception of the election itself as of secondary importance, which maximised transitory anti-government protest voting. However, the Green vote was at its highest in traditional Tory areas and the ethereal nature of this support was far from clear at the time. In the circumstances it was reasonable to believe that Commission criticism of the government's environmental credentials could cause it damage at the next general election. A strategy of contesting such criticisms (rather than admitting any previous environmental laxity) therefore made electoral sense.

The possibility that the furore might die down after an accommodation had been reached on the initial disputes was lessened because the extensive media coverage given to these initial disputes created a snowball effect. More and more information reached the Commission from British citizens and pressure groups as they became aware of the possibility that Commission action could challenge both past and present government policies which threatened the environment. By 1990, one-third of individual complaints made to the Commission about breaches of environmental Directives were coming from Britain, and pressure groups were setting up their own monitoring systems to provide the Commission with information on breaches they could otherwise never have known about. Given that DG XI was more accustomed to EC states providing too little monitoring data with which to identify offences to prosecute, it is hardly surprising that enthusiastic action against Britain should have resulted. The apparent popularity of such action, as shown by Eurobarometer opinion polls, and the opportunity to bring itself closer to the status of the 'big' agriculture and transport directorates no doubt encouraged DG XI officials to continue taking an inflexible line over Britain's compliance with its Directives too.

The Impact of the Compliance Rows on Commission Policy

It is hard to gauge whether Ripa di Meana believed that his inflexibility could improve environmental protection without causing deleterious political effects. If he did, then it must be counted as a misjudgement because in a period where the future direction of the Community was being discussed, actions which emphasised the present existing erosion of the sovereignty of its members were not likely to help those pressing for deeper integration (a course likely to strengthen the role and powers of DG XI). This was particularly true during the detailed negotiations leading up to the drafting and signing of the Maastricht Treaty. This treaty required the unanimous agreement of all twelve member states. In these circumstances it is not surprising that the constant prodding from DG XI should have provoked the British Government into retaliatory action. The provocative di Meana letter calling for a halt to work at Twyford Down, for example, prompted the dispatch of an icy letter to the Commission by Mr. Major, the British Prime Minister, allegedly threatening to block the signing of the Treaty and, more realistically, to rescind Britain's acceptance of the proposed extension of Qualified Majority Voting for EC environmental legislation.[9]

This is an indication that the Maastricht ratification process considerably strengthened the British Government's hand in its disputes with DG XI. The Commission could not risk swelling the ranks of Conservative Euro-sceptics to the point of losing the Treaty, and as a result DG XI was ordered to tone down its actions. The departure of di Meana from the Commission in mid-1992 also lowered the temperature. The government resisted the temptation to take up Delors's suggestion that Community environment policy should be scrapped in favour of national action (on supposed subsidiarity grounds), and the Commission accepted that the planning procedure for the Twford Down road scheme amounted to an impact assessment (Commission, 1992b). European Court action on bathing waters was delayed twice to save Britain embarrassment during its six-month European Presidency in the second half of 1992.[10]

The British government did use the post-Danish referendum discussion of subsidiarity to push its own view of the proper content of European environmental legislation. Although the complaints of Douglas Hurd that the Community should cease intervening in the 'nooks and crannies of national life' seemed to signal a Foreign Office attempt to diminish the EC's role in environmental policy, the DoE, well aware of the need for European-level action to deal with cross-border pollution, was able to bring official policy into agreement with its belief that subsidiarity implies that the Commission has no role in purely national problems, but must take a role where pollution is not contained within national borders. Subsidiarity is not a new concept for European environmental policy (Haigh, 1984b) and Britain's attachment to it has not been seriously disruptive. Most of the 27 items which Britain suggested for repeal or withdrawal at the 1992 Edinburgh European Council were minor (and in many cases had not yet even been adopted by the Commission).[11] Indeed, behind the aggressive rhetoric (for internal Conservative Party consumption) British plans have in general been positive, seeking to address the major current problems of EU environment policy, such as lack of reliable monitoring, patchy enforcement, excessive cost, failure to take new scientific knowledge into account, and lack of consideration of national/local circumstances. Solving these problems is in the British interest as well as the Commission's (although not perhaps to be welcomed by all member states).

In return the Commission has been conciliatory. It has recognised that many of Britain's complaints have a basis of reasonableness. It has attempted to use the debate on subsidiarity in a positive way to set out new ground rules for the extent of EC legislation and national freedom of action. Its view of subsidiarity is that through its operation the Commission will 'do less but do it better' (Commission, 1992a) This implies a stricter adherence in the future to legislating mainly for problems with a cross-border nature (or to prevent the growth of non-tariff trade barriers) but an increase in the enforcement of compliance. For example, a recent Commission policy paper announced that, 'In line with the subsidiarity principle, the Commission intends to reorientate rules and regulations towards compliance with essential quality and health parameters, leaving member states free to add secondary parameters if they see fit' (Com[93]545). This would seem to suggest that revised environmental Directives will be reduced in scope and cost. It might, for example, be decided that no EU rules on the colour of drinking water are necessary where 'discoloured' water presents no threat to human health. In some cases the use of zero limits for harmful substances might be altered to take into account evidence that low concentrations of certain substances are not dangerous and are very expensive to eradicate. Such moves would lower the costs involved in compliance. But it is unlikely that environmental Directives will cost less to implement if they are revised upwards. It has been suggested, for example, that the proposed revisions to the bathing water Directive could cost Britain up to £1 billion to implement.[12] These revisions, justified on health grounds, are not surprisingly under fire from the water companies for being too strict, but there is also strong pressure from marine conservationists and groups representing recreational users of bathing waters (e.g., surfers), backed up by some scientists, for still stricter limits on pollution to protect their health and the marine environment.

Updating standards is therefore as likely to lead to tougher and more expensive standards as laxer and cheaper ones. This will not, however, necessarily lead to a return to hostile DG XI-British Government relations because there are a number of other factors that seem to lessen this possibility. There is, for example, a recognition that attention needs to be paid to monitoring if Directives are to be enforced more fairly. The rash of complaints against Britain in the late 1980s did not indicate that Britain was 'the dirty man of Europe': rather they reflected the high level of public awareness in Britain about the environment, and the lack of opportunities that the centralised political system in Britain offered to protesters; it did not take long for the Commission to come to be regarded as a very effective resource that domestic campaigners could mobilise against governmental environmental and development policies. The ability of the Commission to take action was also linked to the quality of British monitoring. Speaking in 1990, Ludwig Kramer confessed that, 'We have 47 pieces of environment legislation which require countries to report the situation to us. My estimate on compliance with that is about 25% ... [but] the UK and Denmark [are best].'[13] The fledgeling European Environment Agency could improve this situation and remove the UK's vulnerability to being singled out by information-starved DG XI officials. Other EU states are likely to be revealed as more serious transgressors of Directives.

EFFECTS OF THE EC/EU ON BRITISH DOMESTIC POLICY MAKING, LEGISLATION AND ENFORCEMENT

Effects on the Style of British Policy Making

The addition of a Community dimension to environment policy has radically changed the evolution of British policy making. Traditional British practice could be characterised by a number of features, including:

(a)   reliance on scientific and expert consensus;
(b)   use of minimal framework legislation;
(c)   detailed implementation and standard setting left to local authorities and semi-autonomous special government agencies.

The arrival of Community action has forced changes either in the practices themselves or the outcomes issuing from them.

The tendency to plan environmental policy by deliberations between DoE civil servants and the representatives of approved 'insider' groups, and scientists previously led to much 'non-decision making' (Hill, Aaronovitch and Baldock, 1994) or to consensual, voluntary 'agreements' to lower pollution between government and polluters. Now the existence of proposals for Directives accelerates the arrival of their subject on to the political agenda, even where the problem (as with nitrate pollution of drinking water) is relatively invisible and so not the source of public disquiet and pressure. Once agreed, the existence of standards in the Directives blocks the possibility of 'solving' problems by relaxing the legal or health standard at which the type of pollution is deemed to be unacceptable. It also increases the influence that pressure groups can have, as the EU position legitimises and strengthens the public perception of pressure group claims and policy prescriptions.

The reluctance to introduce new parliamentary Bills to implement new Directives has, however been relatively unaffected. Although the Commission has forced the government to use Statutory Instruments rather than Departmental circulars, the amount of parliamentary time needed to implement Directives by this method is still small. It has also invested existing bodies, such as the Health and Safety Executive, with responsibility to expand their work to include environmental considerations, thus avoiding the time and expense of setting up new monitoring agencies, except when absolutely necessary (e.g., the NRA).

This apparent continuity of style, however, hides the extent to which the content of environmental legislation has been determined at the European level and the implementation of environmental standards has been further centralised in the UK.

Where new Acts of Parliament have been necessary, these have taken into account the obligations set by existing Directives together with the obligations of those likely to be incurred by those still passing through the EC/EU legislative system. Both the Water Acts (that of 1989 and the 'consolidated' acts of 1991) and the Environmental Protection Act 1990 reflect decisions taken at a European, not domestic level. The large numbers of Statutory Instruments made in order to implement environmental Directives have become increasingly detailed in their desire to implement Directives to the letter and avoid enforcement action by DG XI.[14]

Centralisation in UK environmental enforcement is directly linked to the increasing decisions on policy content at the European level. All EU legislation is agreed by national governments in the Council of Ministers, and so all aspects of the environment covered by Directives become national matters. The UK central government is responsible for any failures to achieve Directive standards and so the autonomy of local authorities and regional or national non-governmental regulators has been much reduced (Haigh, 1986). The situation in the early 1980s in which a few urban councils regarded smoke control as a low priority could not be allowed to continue once the Commission could drag the government (not the councils) into the Court of Justice to explain why Directive standards had been broken. Long unused powers allowing the government to force councils to meet standards have been activated: an example is the use in the mid-1980s of the government's dormant powers under the 1968 Clean Air Act to order Sunderland and Wakefield councils to institute smoke control programmes. The 1990 Environmental Protection Act extended these powers over local councils to the control of other forms of air pollution. The authority of pollution regulators to interpret legislation and define acceptable levels of pollution has also been curtailed for the same reason.

Another change is that the technical nature of Directives has altered the content of the issues over which ministers and civil servants must negotiate. Traditionally, the setting of technical standards and their enforcement had been devolved to bodies such as the Water Authorities and HMIP. A Union of 12 sovereign states, however, has found itself driven to setting exact numerical standards if the desired results are to be achieved. Experience has shown that the Commission is powerless to force recalcitrant states to bring about the aims of a Directive if there are no numerical standards or specific deadlines fixed. If the environment is to be protected from cross-border pollution and the single market from environmentally 'justified' barriers to free trade, the supranational enforcement that the EU can provide is essential. It must be clear what everyone has signed up to and when legal action by the Court of Justice is warranted. As a consequence of this British ministers and civil servants now need to be able to argue out technical matters in Brussels as new Directives are negotiated. These negotiations may, however, become easier as the difficulty of setting rigid standards that are appropriate for all parts of the EU's geographical area becomes increasingly clear. The Council and Commission will probably seek to incorporate more flexibility in Directives.

Independent Policy Initiatives

Britain has not lost all opportunity for developing initiatives in the environmental field. Firstly, as one of the biggest and most influential members of the Community it can choose to influence the discussions in the Council of Ministers (and unofficially in the Commission too) in positive as well as negative ways. As we noted in the introduction, Britain was at the forefront of the process leading to the elimination of lead from petrol, and more recently it has been among those pressing for the establishment of a European Inspectorate to monitor pollution. In this latter case Britain has had to be content with the establishment of the European Environment Agency which will have fewer teeth than an Inspectorate, but will be able to improve the flow of information on which the Commission bases its enforcement action.

There have also been domestic initiatives designed to prevent decisions by other government departments from undermining environmental objectives. The *This Common Inheritance* White Paper of 1990 aimed to integrate environmental concerns into all government decision taking. Departments list their commitments to environmental protection and appoint one of their ministerial team as 'green minister', responsible for putting these commitments into practice. The DoE also monitors the making and fulfilment of these pledges. The government issues an annual report detailing steps taken. This whole procedure may well not work in practice as well as it is designed in theory. The Department of Transport seems to have a rather different view of environmental protection from the DoE and there have been claims that the annual government reports have become patchy, opaque and secretive. However, in the context of this discussion, it is the fact that the whole initiative is a domestic one, not a direct response to an EC Directive that is important. Moreover, it is certainly possible to argue that the existence of EC legislation was not a decisive influence on the *This Common Inheritance* strategy in general.

This British initiative has influenced the future workings of the Commission, as it seeks to integrate environmental considerations into its other policies. Each Directorate General has recently designated a senior official to ensure that environmental implications are taken into account in proposed Directives and Regulations and there are to be assessments published of progress made towards the aims of the current Environmental Action Plan. As with the British DoE there is also an internal DG XI section dedicated to monitoring the environmental performance of other directorates.

The Future Effects of the EU on British Environmental Policy

The future effects on Britain of EU environmental policy are hard to gauge. This is because the several pointers to the future are facing in different ways. As noted earlier, it is far from clear that the revision of environmental directives, even with the new found emphasis on subsidiarity, will result in laxer standards which are easier to comply with or legislate for. However, the current political and economic realities in the EU, the current emphasis on even-handed monitoring and enforcement, and the new attention paid to the cost of measures and the desirability of using more flexible voluntary arrangements to tackle problems may suggest a period of harmony, with Britain exerting much positive influence.

The effects on Britain of the ratification of the Maastricht Treaty with its revised rules for EU environmental legislation are far from clear. There are a number of changes which will effect the process of EU environmental legislation.[15] Measures made under Article 100a (the 'single market' section of the Treaty which is often used for environmental objectives to prevent different standards becoming a barrier to trade) will continue to be subject to QMV in the Council, but the European Parliament now has a 'co-decision' rather than merely a co-operation role. This essentially means that MEPs are able to veto such legislation if they consider that not enough of their amendments have been accepted. Agreement on the next Environment Action Programme will also take place in this way (Art. 130s(3)). Most environmental measures not brought in under 100a will be under 130s(1) which moves from the previous system of unanimous Council votes with mere consultation of the Parliament to QMV in the Council and the cooperation procedure in the Parliament (which gives MEPs a chance of making amendments). A rather vaguely defined range of areas including primarily fiscal measures, water resources, domestic energy supply and land use/planning retain the old system of unanimity in the Council and a consultative role for the Parliament (130s(2)).

The vagueness of the Treaty means that there are likely to be many dis-agreements between the Union institutions over which legal basis and which process should be followed for new items of legislation. This in itself may well result in a reduction in the number of new Directives that Britain has to cope with in the short term. The extensive involvement of the European Parliament in Article.100a legislation will delay the arrival of such legislation, but may also lead to the inclusion of measures which the British Government may dislike. The relative weakness of Conservative representation in the Parliament will reduce the current British Govern-ment's influence there and, with the exception of those measures falling under Article 130s(2), the British Government cannot veto proposals without the aid of other countries. This might signal a period of renewed Britain-Europe conflict but there are other factors at work which either make this unlikely or else offer no certainty of renewed conflict..

First, the rule changes contained in the Maastricht Treaty have removed one of the main reasons encouraging DG XI and the European Parliament to work closely together in the Single European Act era. Previously it suited both to try to introduce legislation under 100a which maximised the powers of the Parliament and removed the blocking veto of individual states which still applied to 130s legislation. Now that QMV is the norm for almost all environmental matters the Commission and the Council may now wish to cooperate to use Article 130s and thus avoid giving Parliament the powerful (and time-consuming) say provided by the co-decision procedure under 100a (Wilkinson, 1992).

Second, the extension of QMV to most environmental matters will result in a few more decisions being taken against British consent as the minority in these matters tends to consist of Denmark, the Netherlands and Germany who wish to proceed further or faster than the other states. The current economic problems and government budget deficits across the Union may also increase support for the British Government's view that high standards can be achieved in a cost-effective way by a careful and gradual approach. However, it is always possible that even a modest increase in the number of Directives approved against British opposition could become a renewed source of conflict between the Union and the British Government. Reaction to the overruling of British opposition to proposals to outlaw the mixing of domestic and hazardous waste in landfill disposal sites may be illuminating in this regard.

Finally, the current political climate in Europe may make the Commission, and possibly even the European Parliament, take extra care not to take steps which they know could seriously antagonise member states, particularly large and powerful member states such as Britain. This political climate originates from the shock that the European political establishment received from the near failure of the Maastricht ratification process; the declining turnouts and high levels of support given to some anti-Union parties in the 1994 European elections; and from the long 1990s recession which has reduced the priority given to environmental matters by governments and the mass public. The economic problems in Europe have also reduced the willingness of countries such as Germany (which in addition now has to divert resources to the former East Germany) and France to agree to rapid introductions of strict environmental standards if these can only be implemented by subsidies to the poorer members of the Union. This is likely to make the Commission think more about the costs of its proposals although, as noted earlier, the proposals for a revised bathing water Directive do not go very far in this direction. Whether the member states will demand further relaxation through the Council of Ministers remains to be seen, but if they do DG XI will be weakened in the impending negotiations over its future powers at the 1996 Intergovernmental Conference: it will not want to make too many enemies in the run up to this vital conference.

This suggests a period of relative peace between DG XI and the British Government. If, as seems likely, DG XI puts much of its energies into revising and monitoring the Directives already in place, then Britain (which has done much to meet those already in force) is likely to have a relatively quiet time, whereas others such as Italy, Greece and Belgium are increasingly likely to be put under pressure as improved monitoring reveals the extent of their non-compliance with existing Directives.

We must also note that the extent to which EU Directives will affect British environmental policy is increasingly in the hands of the British

Government itself. Britain holds the median position (between the Danes, Dutch and Germans on the one hand and Ireland and the Mediterranean countries on the other) and so is in a potentially influential position. How the government uses this influence depends to a large extent on how it answers two questions: (a) how far is it willing to see the growth of non-tariff barriers to the single market for the sake of nationalising environmental policy, and (b) how willing is it to accept that some problems affecting Britain's environment need action across borders? The DoE (if not the euro-sceptic wing of the Conservative Party) seems to have come to a view, based on a common-sense reading of subsidiarity, that a distinction can be made between measures which tackle cross-border pollution or the erection of non-tariff trade barriers and those which have no effect beyond the domestic level, or no effect on tradable goods. The first type of measures is to be supported; the second type is to be avoided. In time it should be possible to see if this strategy is being followed consistently within the EU itself.

CONCLUSION

We must not lose sight of the massive shift in power from national states to the EU which has occurred in environmental policy in recent years. This shift has taken place at the same time as the environment itself has become a more important issue and, in some cases such as the ozone layer, an urgent priority. Britain could not have isolated its environmental policy from these wider developments, and the existence of agreements like the Montreal Protocol on CFCs shows that a purely domestic environmental agenda is an impossibility in the world today.

Signing international agreements presents much less of a challenge to British autonomy over its own environmental policy than do its EU obligations. International agreements do not contain the element of supranationality and shared sovereignty that the EU with its Commission and Court possesses. This supranationality makes it inevitable that there are occasions when the British Government is forced to impose environmental standards more quickly and strictly than it would like. The form of this action is also restricted by Community supranationality. The quarrels of 1989–92 may be seen in retrospect as the Commission and the British Government respectively testing the limits of their power and influence as a precursor to establishing a new equilibrium. This process might have been less painful if a different combination of personalities had been present, but given the requirements of the Directives in question some form of limit-testing and disequilibrium was inevitable.

Notes

1. To what extent the removal of lead from petrol was the result of the campaign and to what extent it was the result of a Royal Commission cannot be determined. For a discussion, see the chapters by Sir Richard Southwood and Nigel Haigh in Conway (1986).

2. A letter from Nigel Haigh published in the *The Times* on 11 February 1982 discussed the EC Directive and concluded, 'It follows that for CLEAR to be successful, it will have to take its campaign to several European countries simultaneously.'

3. Paper to a conference of the Royal Institute of Public Administration, 21 September 1990.

4. Working document of the European Parliament 1-109/84, 9 April 1984, 'Report drawn up on behalf of the Committee of Enquiry into the Treatment of Toxic and Dangerous Substances by the European Community and its Member States'.

5. Official Journal, C 127, 14 May 1984.

6. A summary of Ludwig Kramer's beliefs in the role of environmental lawyers can be found in Kramer (1989).

7. Note that in the poorer EU states there tends not to be high levels of mass public concern over environmental standards and it is accepted that the standards protected in domestic law will be only those laid down by the EU. There is also no self-image of being the 'good guys' who fulfil their commitments while others dodge theirs (not least because some of these countries, along with Belgium, are often the worst at complying with Directives).

8. *The Independent* , 22 October 1991, 1.

9. Ibid.

10. *Financial Times*, 1 July 1992, 2.

11. *Financial Times*, 8 December 1992, 2.

12. *ENDS Report*, 234 (July 1994), 17–21.

13. *The Independent,* 8 May 1990, 8.

14. The example of the Waste Management Regulations 1994, succinctly satirised in the leader article of *Waste Manager*, June 1994, shows the extent to which Statutory Instruments are often largely transpositions from Directives.

15. The following paragraph draws on the analysis provided by Wilkinson (1992).

References

Commission of the European Communities (1992a), 'The Subsidiarity Principle' *Bulletin of the European Communities,* 10-1992, 116–26.

Commission of the European Communities (1992b), *Press Release* (31 July 1992 *United Kingdom infringements – The Commission terminates five procedures and sends two reasoned opinions.*

Commission of the European Communities (COM [93] 545), *Commission Report to the European Council on the Adaptation of Community Legislation to the Subsidiarity Principle.*

Conway, G. (ed.) (1986), The Assessment of Environmental Problems, Imperial College Centre for Environmental Technology.

Haigh, N. (1984a), *EEC Environmental Policy and Britain: An Essay and a Handbook* (London: Environmental Data Services, or ENDS).

Haigh, N. (1984b), 'The Environment as a Test Case for Subsidiarity': paper delivered at 'Maastricht and Subsidiarity and the Environment' Conference held at the London School of Economics, *Environmental Liability*, 2.

Haigh, N. (1986), 'Devolved Responsibility and Centralization: Effects of EEC Environmental Policy', *Public Administration*, 64, 197–207.

Haigh, N. (n.d.), *Manual of Environmental Policy: the EC and Britain*, looseleaf (London: Longman).

Hill, M., Aaronovitch, S. and Baldock, D. (1994), 'Non-Decision Making in Pollution Control in Britain: Nitrate Pollution, the EEC Drinking Water Directive and Agriculture', *Policy and Politics*, 227–40.

Johnson, S. P. (1979), *The Pollution Control Policy of the European Communities*, 2nd edn 1983 (London: Graham & Trotman).

Kramer, L. (1989), 'The Open Society, its Lawyers and its Environment', *Journal of Environmental Law*, 1, 1–9.

This Common Inheritance: Britain's Environmental Strategy (1990) (London: HMSO) Cmnd 1200.

Wilkinson, D. (1992), 'Maastricht and the Environment', *Journal of Environmental Law*, 4, 221–39.

4 The Establishment of a Cross-Sector Environment Agency

Neil Carter and Philip Lowe

INTRODUCTION

> We will establish a new Environment Agency which will bring together the functions of the National Rivers Authority, Her Majesty's Inspectorate of Pollution and the waste regulation functions of local authorities. (*Conservative Party Manifesto 1992*)

The promise to set up a new cross-sectoral environment agency was the centrepiece of the environment policy section of the Conservative Party manifesto for the 1992 general election; to be precise, it was the only domestic environmental policy pledge in the manifesto. The pledge, in part, reflected a remarkable near-consensus across government, opposition parties and the network of pressure groups with an interest in the issue of environmental regulation. This agreement was remarkable on two counts. First, the creation of an agency with cross-media regulatory responsibilities for air, water and land would appear to represent a significant departure from the traditional pattern of British environmental administration. Second, rarely is there agreement across such a range of producer, consumer and environmentalist interests from different policy networks. It is therefore not surprising that the political consensus stretches only to the principle of establishing such an agency, and that there has been considerable disagreement both within government and between pressure groups as to the form and functions it should take. This chapter examines the thinking behind the proposed agency because it provides a useful insight into the British administrative system and the pressures to which it has been subjected in recent years.

THE POLITICAL CONTEXT

John Major announced the Government's intention to create a combined environmental agency on 8 July 1991. Yet just ten months earlier, the

White Paper, *This Common Inheritance*, had rejected change in favour of allowing the newly-formed unified pollution inspectorate HMIP, the NRA, and the unified conservation agencies for Scotland and for Wales (Scottish Natural Heritage and the Countryside Council for Wales respectively) to establish themselves.

Major's change of heart was undoubtedly motivated by immediate political considerations. It was his first speech on the environment as Prime Minister and he was keen to establish his own 'green' credentials in a way that his predecessor, Margaret Thatcher, had been unable to do. In the run up to the general election, with both opposition parties in favour of the need for institutional reform in the environmental field, he was also intent on minimising the possibility of party competition over this issue (Carter, 1992). Indeed, later on the same day as Major's speech, the Labour Party was due to publicise a new Fabian pamphlet setting out its own plans for an environment agency.

However, although there was general (if sometimes lukewarm) support for an agency, the details of the proposal generated bitter conflict within Whitehall as MAFF, represented by John Gummer, Minister of Agriculture, fought to secure control over various river authority functions which would be left over if only water pollution regulation was to be absorbed by the new agency. Fearful of the consequences of dismembering the newly-formed NRA with its growing reputation for 'taking on' producer interests, the environmental lobby mobilised in its defence. This conflict tarnished the Prime Minister's personal initiative. Rather than risk the dispute simmering over in the lead up to the general election, Michael Heseltine, Secretary of State for the Environment, attempted in October 1991 to defuse the controversy by publishing a consultation paper setting out a number of options for the new agency (Department of Environment, 1991). The period of consultation would last until January 1992 which would delay any final decision on the form of the agency until well beyond the date of the election.

Since then the omission of an environment bill from the Goverment's legislative programme for the 1992–93 Parliamentary sessions had prompted speculation that the proposal would be quietly dropped. Although Michael Howard, Secretary of State for the Environment, reiterated the Government's commitment to the agency in July 1992 neither the immediate political origins of the agency proposal – characterised by political symbolism, transitory leadership commitment and the disappearance of the issue during the general election campaign – nor the subsequent decline in the political significance of the environment augured well for the proposal agency. Eventually, however, an Environment Bill was introduced during the 1994–5 parliamentary session with the intention of setting up the agency by the end of 1995.

THE EMERGING CONSENSUS

It was always unlikely that the Government would ditch the project. John Major's announcement was not merely political opportunism; it also reflected the culmination of a process of rethinking Government policy that had begun at least two years beforehand.

Early in 1989 Nicholas Ridley, then Secretary of State for the Environment, had rejected a recommendation by the House of Commons Select Committee on the Environment to set up an inclusive agency, arguing dismissively that 'there is no guarantee that synergy will be gained automatically by amalgamating disparate organisations' (ENDS Report No.198). Nevertheless there was considerable and growing support for an inclusive environmental agency in Government circles. William Waldegrave, when Minister of State for the Environment, had devoted a speech to the issue as far back as October 1984. The Conservative-controlled House of Commons Select Committee on the Environment had supported the concept in 1989. Chris Patten considered it when drawing up the White Paper during 1989–90 and, although *This Common Inheritance* rejected immediate amalgamation of HMIP and the NRA, it did suggest that in the long term an umbrella body might be set up to oversee and coordinate their activities. David Trippier, Minister for the Environment and Countryside, told the Environment Committee that the Government was 'very seriously considering' an agency two months before Major's announcement. As the House of Commons Environment Committee (1991–2) put it: 'In just two years, the Government's policy had shifted from outright rejection of the notion of such an agency, to enthusiastic acceptance.'

Yet support for an agency sits uneasily alongside the established traditions of environment policy. Britain has the oldest system of environmental protection of any industrialised nation and probably one of the most elaborate. Briefly, five main features characterise this distinctive administrative system and style of policy making.

First, government structures and legislation have been a fragmented accretion of common law, statutes, agencies, procedures and policies. There has been no overall environmental policy other than the sum of the individual elements, most of which have been pragmatic and incremental responses to specific problems and to the evolution of scientific knowledge.

Second, regulation has been highly devolved and decentralised: responsibility for taking action has fallen mostly to local government or to one of a wide range of administrative agencies. The roles allocated to these bodies usually combine policy development and advice with regulation and implementation.

Third, environmental control, like many other aspects of state regulation in Britain, is pervaded by administrative rather than judicial procedures. The approach pursued is informal, accommodative and technocratic rather than formal, confrontational and legalistic. Legislation tends to be broad and discretionary. Regulatory agencies are usually given wide scope to determine environmental objectives and considerable latitude in their enforcement. When laws are demonstrably broken, prosecutions have traditionally been rare because regulatory officials prefer to exercise the considerable discretion allowed to them by their relative insulation from the scrutiny of both parliament and the judiciary.

Fourth, the voluntarist approach adopted in dealing with private concerns seeks to foster cooperation and strives to achieve the objectives of environmental policy through negotiation and persuasion. Where other countries have made extensive use of uniform standards of emissions and environmental quality backed up by law, the British Government prefers implementation through consent by means of industry self-regulation and informal agreement. For 'the British are reluctant to adopt rules and regulations with which they cannot guarantee compliance' (Vogel, 1986: 77). The voluntarist model, based on a commitment to confidentiality between regulator and regulated, thus fits perfectly into the broader British administrative culture of secrecy.

Lastly, environmental policy making where it impinges on major economic interests, like pollution control, has traditionally taken place in closed policy communities in which producers – industrialists, trade associations, farmers and landowners – are strongly represented. Consequently, the negative view of the voluntarist tradition is the belief, widely held amongst environmentalists, that government officials have been 'captured' by industrial managers and agricultural interests.

These core characteristics of British environmental policy remain dominant but, by the late 1980s, pressures from a number of different directions had built up to stimulate a major reappraisal of policy and structures. In particular, support grew for a consolidation of regulatory functions. Much of the intellectual case was formulated by the Royal Commission on Environmental Pollution. For many years it has argued for unified control of discharges to air, water and land, to avoid the all too common situation where control over one media simply displaces the problem to another, without any overall strategy for the most appropriate treatment of disposal routes. This key concept of integrated (or cross-media) pollution control (IPC) was behind the Commission's calls for a unified pollution inspectorate, which eventually came about with the creation of HMIP amalgamating the air, water and solid waste, and radioactive waste directorates. In itself, it was a modest administrative reform motivated primarily by the desire to improve the technical effectiveness of pollution control within the established traditions of British pollution control practices (O'Riordan and Weale, 1989). Nevertheless, soon after, the DoE issued a consultation paper (Department of Environment, 1988) making clear its new commitment to IPC: a decision that, in breaking with established traditions, reflected a general international movement away from 'end-of-pipe' approaches. In the light of its acceptance of IPC and with HMIP still lacking a coherent mandate, the government introduced the Environmental Protection Act 1990 (EPA) which created the framework within which HMIP could operate (although, characteristically, it prescribed a gradualist, step-by-step, implementation of the terms of the Act).

There remain, however, many polluting and hazardous processes that HMIP does not regulate and there are regulations concerning the use of particular media (most notably water) that potentially cross-cut HMIP's powers. Indeed, soon after creating HMIP the government established first, the NRA under the Water Act 1989 (see below) and, then, local Waste Regulation Authorities under the EPA. None of these bodies was formally within the legal framework of integrated pollution control. With respect to the NRA the matter was resolved by transferring primary responsibility for the majority of discharges from it to HMIP. The NRA was left with the power to set conditions relating to discharges to controlled waters within the authorisation given by the HMIP. As regards Waste Regulation Authorities (which are local authority based), the confusion remained unresolved. The Environment White Paper put forward the possibility of giving HMIP the leading responsibility for the regulation of waste disposal (under the EPA it already has the duty of auditing the performance of local authorities in this field). Significantly, though, when the Opposition moved an amendment in the Lords to the EPA in June 1990 which would have achieved just this resolution, Lord Hesketh, a Government Minister, rejected it with the comment, 'Local knowledge and the capacity for a fast response which lies with local authorities could be put at risk by a centralised system.'

Clearly, rather than resolving the issue of the consolidation of regulatory functions, the formation of HMIP simply fuelled the debate in environmental circles. Recognition of such divisions has prompted calls for further rationalisation of the structure of regulation. For O'Riordan and Weale (1992), they are examples of 'inappropriate coordination and mismanagement of what should be a total environmental regulatory function'. The arguments for amalgamating regulatory bodies are summarised succinctly in *This Common Inheritance*:

> It might make it easier for industry and others to have only one inspectorate to approach; that with separate organizations there is always the risk of some overlap or duplication of effort; and that combining the bodies under one management might lead to greater consistency of approach across all pollution types and environmental media (para.18.13)

Support for this position grew rapidly. Submissions to the consultation paper reveal support for an encompassing regulatory agency from a broad coalition of interests ranging from the Confederation of British Industry (CBI) to most environmental groups.

The Government's response to the call for a consolidated Environmental Protection Commission made by the Select Committee on the Environment indicates one of the potential counter-arguments. The response, published in April 1989, rejected the proposal on the grounds that a multiplicity of agencies, tailor-made to regulate specific issues, was more effective.

However by 1991 things had moved on rapidly and the new Secretary of State, Michael Heseltine, was able to 'look at these matters afresh' (House of Commons Environment Committee, 1991–2: 1). What he observed was a widespread, growing consensus in environmental circles about the need for some kind of cross-sector agency, although the form that the agency should take remained a matter of dispute.

WHAT FORM OF AGENCY?

The decision to issue a consultation paper in the autumn of 1991 reflected the existence of substantial divisions, both inside and outside government, regarding the best form for any future agency. Thus the consultation paper set out four possible options and provoked over 600 responses. Never before had there been such a welter of proposals for new official structures for environmental protection. This spate of institutional blueprints marked a significant departure not only in indicating a gathering consensus over the need for major reform, but also in breaking with the strong tradition of institutional pragmatism in Britain whereby machinery and measures were devised to address the problem to hand rather than by reference to more general principles. Nevertheless, to raise the chances of successful reform of the system of environmental protection, proposals had to recognise the nature and structure of existing agencies and procedures, and work 'with the grain' of the British administrative tradition.

A number of issues have come to the fore in the reform debate: the degree of independence that the regulatory agencies should enjoy; the role of local authorities; the scope for and limits to the consolidation of regulatory functions; and the desirability of and scope for separating operational roles from regulation. The dilemmas involved and their resolution go to the very heart of the British tradition of environmental management.

The arguments for independence from government seem very persuasive. Indeed, some commentators take the case as self-evident. O'Riordan and Weale, for example, simply declare that 'independence of regulation both from those who are licensed and from governments that might interfere' (1992, para. 2.11) is a fundamental principle on which the greening of the machinery must be founded. More explicitly, the Select Committee on the Environment, following its investigation of toxic waste, proposed the need for an independent regulatory and enforcement agency in the light of the scandalous dilatoriness of the DoE and local authorities in implementing the requirements of the 1974 *Control of Pollution Act* (House of Commons Environment Committee, 1988–9). Environmental protection, it was felt, should not be so subject to the vagaries of political fortune. To maintain public and scientific confidence demanded objectivity. The case for independence has also been put in terms of avoidance of possible conflict of interest problems where a Minister has to settle appeals. The final argument for not putting a regulatory agency under the control of any particular department is that the agency's responsibilities should not be restricted by departmental boundaries.

The counter-argument for retaining regulatory agencies as departmental bodies is that this ensures political accountability (a principle underpinning the Next Steps agency initiative). Ministers argued against various proposals for an Environmental Protection Executive during the passage of the EPA in just these terms. It was claimed that to devolve important responsibilities to an outside statutory body would remove them from the normal system of parliamentary control and scrutiny. However, such an arrangement is not unusual and the so-called Addison Rules, which relate to parliamentary scrutiny in these cases, specify that when questions affecting public bodies arise in Parliament, the Government is responsible to Parliament alone. In practice, however, Ministers often refer parliamentary questions to the head of the appropriate body to answer by correspondence.

In the event, the Government has decided that the agency should be independent of Government. Members of its Board will be appointed by Government, presumably with a mix of industrialists, agriculturalists, environmentalists and scientists. It will be answerable to Parliament either directly or via a minister. This arrangement does not entirely resolve another long-standing accountability issue relating to local government. There are two extreme models of the local role of the proposed agency: one view envisages it maintaining consistency across a series of locally organised activities; the other sees it as a single national body responsible for regulation as well as strategic planning. Currently it is envisaged that the agency will fall somewhere between the two perspectives, raising the problem of how, for functions remaining with local authorities, the agency can be accountable simultaneously to Parliament and to locally elected officials.

Local authorities have attracted criticism for their record in waste management, particularly in planning sufficient disposal capacity and in establishing handling and treatment facilities for specialised wastes. The EPA sought to address the problem by separating the operational and regulatory functions with the intention of opening up waste disposal to competitive tendering and encouraging local authorities to combine in regional waste disposal companies operating at arms length from the regulatory authorities. However, considerable unease remained concerning the effectiveness of this voluntary strategy to overcome the tendency of local authorities to 'free ride' by hoping that neighbouring authorities would accept their waste. Consequently, all options outlined in the consultation paper envisaged removing responsibility for waste management regulation from local authorities and vesting it with the new agency. The central argument for this approach is the possibility of developing a national waste regulation strategy, and making best use of resources and expertise that are currently distributed inefficiently among a multitude of authorities. In addition, a single agency will create the 'one stop shop' organisation so attractive to industry.

This centralisation of powers away from local government follows a pattern that the Conservative Government has adopted in other policy areas. Local authority associations employed fashionable concepts in their defence. The principle of subsidiarity was cited to emphasise local expertise, knowledge and speed of response against the putative benefits of greater managerial and technical competence claimed for centralisation. The 'democratic deficit' was deployed to condemn the proposed agency for lacking the accountability of elected local members. They also argued that the new organisations arising out of the EPA had not yet had the opportunity to prove themselves. But the Government had, apparently, already made up its mind. After all, subsidiarity had resulted in the very problems that now need solving; other regulatory bodies (such as the NRA) were not directly accountable to elected representatives; and once the principle of an agency had been accepted, then if the newly-established HMIP and NRA were to be re-organised, why not waste regulation? However, it is unlikely that the new agency will be rid of the local contention that surrounds the siting of waste management facilities.

The issue that has probably caused most controversy concerns the degree of comprehensiveness of possible regroupings: that is how inclusive should the new agency be? The more all-embracing the proposed regrouping, the more likely that it produces a large, unwieldy structure and that the upheaval and difficulties involved in the re-organisation will outweigh, at least in the short term, any advantages to be gained. Even in the longer term, it is possible that existing problems of external coordination would be reproduced as difficulties of internal coordination. The need, for example, for the sort of memorandum of agreement that the HMIP and NRA have had to negotiate is hardly likely to disappear just because the two organisations have been subsumed into the one agency. On the other hand, although a less comprehensive regrouping might reduce some of these difficulties, other problems would arise, particularly in determining the boundary between what should be included and excluded, and then ensuring effective liaison and coordination across that boundary. Significantly, the different proposals for an environmental agency each contained somewhat different groupings of functions. While there was a broad agreement that HMIP should be central to any new agency, it is the fate of the NRA that has divided opinion. Some proposals incorporated it; others excluded it, others would have dismembered it.

The consultation paper presented four options for institutional reorganisation. Under the first option, the new agency would have had a remit for emissions to air and land by taking over responsibility for most of HMIP's functions, together with those of waste regulation from local authorities. HMIP's responsibilities for the water environment would be transferred to the NRA. A second option was to have created an umbrella body to coordinate the work of the NRA and HMIP. The third option was to incorporate HMIP, waste regulation and the NRA within a single, all-embracing agency. The final option was for the agency to combine waste regulation, HMIP and the NRA's pollution control functions, thus consolidating responsibilities for emissions to air, land and water, but leaving a rump of NRA activities covering river, water and fisheries management. For each option, the relative merits and drawbacks were rehearsed, but (as the consultation paper candidly admitted) 'the Government has not reached a firm view on which of these options might best fulfil its objectives for the environment agency' (Department of Environment, 1991, para. 43).

Close examination reveals that two quite different notions of integration were being played off against one another, one being intra-sectoral, the other inter-sectoral. The inter-sectoral notion, which is central to the concept of IPC, is about the pursuit of a coordinated and coherent strategy of environmental protection across different sectors and media.

The NRA, in contrast, is a clear example of a multi-purpose agency based on the integrated management of a single resource. As Lord Crickhowell, the Chairman of the NRA, has commented:

> It is ... a waste of time talking in grandiloquent, generalised terms about environmental protection agencies, without stopping to define objectives or to examine the nature of the organisations which might be incorporated in any such authority. The NRA, for example, is not simply a regulator. It is also manager of a major resource – water. I believe that it would be a profound mistake to dismember that organisation and separate its functions (*House of Lords Debates*, 18 May 1990, col. 526)

The intra-sectoral notion is central to the idea of sustainable development and involves reconciling conflicting demands on particular natural resources. It is this sense of the term that is meant, for example, when the Government discloses in the White Paper that one of its objectives is 'to integrate agricultural and environmental policy' (para. 2.7).

This raises another area where the principles guiding regulatory reform need to be clarified; namely the separation of operational roles from regulation (an issue central to the debate about reorganising waste regulation). The key aim here is to reduce the potential scope for 'gamekeeper-turned-poacher' conflicts of interest. This argument was used by those in favour of fragmenting the NRA which was accused of both regulating pollution and managing the river basin. So, for example, who would bring the NRA to task when, in seeking to raise revenue, it damages the environment by drawing too heavily on its navigation or abstraction functions? The counter-argument underlined the merits of integrated river basin management and stressed that the NRA's pollution control functions cannot easily be separated from its management functions. In short, it is impossible to control pollution without exercising control over water volume. To prevent the gamekeeper-poacher conflict it is not necessary to separate control or enforcement roles from promotional roles, or to strip from regulatory authorities all their other executive functions. That would be to take an unnecessarily narrow view of regulation and diminish the scope for a 'sticks-and-carrots' strategy or for the pursuit of preventative policies.

Quite clearly, divisions over re-organisation go deeper than this debate about abstract principles of form and function: they reflect the views of different interests and the way those interests are responding to the changing arena of environmental policy making. The various approaches attracted the backing of different powerful interest groups, although the majority preferred either option three or four: that is, they polarised around the question of whether the NRA would be transferred wholesale into a new agency or whether its regulatory and operational functions should be fragmented. Industrial interests, led by the CBI plus some local authorities, favoured the notion of a unified environmental agency centred on the HMIP. Business leaders won backing from the influential Environment Select Committee. One of the arguments used was that business needs a one-stop agency for all of its licensing needs. Traditionally, also, firms have had close links with the pollution inspectorate (dating back to its former life as the Alkali Inspectorate), whose officers usually have a training in process or production engineering and practical experience of the industry they are regulating. The cooperative, non-interventionist relationship with industry attracted accusations that the inspectorate had been captured by those interests that it should be regulating; the HMIP has sought to throw off this image by demonstrating that it is no pushover (Jordan, 1993).

On the other hand, the NRA drew strong support from fisheries interests, farmers, riparian owners and amenity interests. They argued the need for a single, powerful agency to mediate the various demands on the water within river catchments to ensure the healthy state of the river and its associated fish stocks and aquatic habitats. These interests found particular support amongst backbench members of Parliament, particularly on the Conservative side. Adding weight to these arguments has been the growing reputation of the NRA as a more interventionist and pro-environmental agency.

Nevertheless divisions appeared on the issue within the environmental movement. Friends of the Earth has been a keen advocate of a unified environmental agency, but traditional nature conservation and countryside interests were initially alarmed at the possibility that the NRA might be dismembered or emasculated within such an arrangement. There was particular concern lest, shorn of its water pollution responsibilities, the rump of the NRA's responsibilities for water and fisheries management and land drainage were to be absorbed by the Ministry of Agriculture (fears which the press suggested were not without foundation). But a common position was reached across the environment lobby to accept a new grouping of regulatory functions as long as this did not involve any dismemberment of the NRA.

Divisions also appeared between government departments. As we have seen, Heseltine's decision to issue a consultation paper in 1991 was prompted partly by a fierce internal row between the DoE and MAFF over the redistribution of NRA functions. Since then this inter-departmental tension has not been helped by recent official interest in the overall responsibilities for coastal protection and sea defence. Primary responsibility for these matters currently rests with MAFF. Operational responsibilities for flood defence reside with the NRA, and for coastal protection with maritime district councils. In 1992 the National Audit Office (NAO) published a report on coastal defences in England that was critical of the priorities and organisational responsibilities in this field (NAO, 1992). A subsequent report by the Environment Committee was also critical of the lack of consideration of environmental factors in coastal defence policy. The Select Committee argued that it was no longer appropriate for coastal defence policy to rest with MAFF. Instead, the DoE would provide a more suitable lead on coastal defence issues, operating through the NRA. In its response, the Government rejected the charge that the allocation of ministerial responsibilities had resulted in any lessening of its commitment to the environment (Department of Environment, 1992). Nonetheless it did announce that an inter-departmental group to coordinate coastal policy had been established, and was coordinated through the DoE's directorate of rural affairs.

In the event, Michael Howard declared the Government's preference for option three: to form an agency which would amalgamate the functions of the NRA, HMIP and the waste regulation work of local authorities. The Agency will cover England and Wales and a separate body will be set up for Scotland. However, the debate itself is not foreclosed by the Government's decision.

DISCUSSION

The circumstances surrounding the decision to form an environment agency suggest that there is considerable instability around some of the established traditions of environmental policy in Britain. Of the five central characteristics of the administrative system and style of policy making outlined earlier, there is particular evidence of change in two. First, the formation of a cross-sector environment agency would break with the tradition of a devolved and decentralised regulatory structure. Second, the decision to retain all the functions of the NRA within the new agency suggests that the strength of the closed pollution policy making community is weakening. In this concluding section we explore these developments.

There are important implications arising from the shift away from devolved and decentralised regulation. Environmental protection is not a traditional self-contained policy sector like agriculture, energy, transport and industry. Ideally it should cut across all these and act as a guiding principle within each. Even the Royal Commission has recognised, in the past, that 'it would be unrealistic to expect one Department of State to be responsible for all aspects of pollution and that there are strong arguments for requiring environmental protection to be an integral part of good management in all areas of Government activity' (Seventh Report, 1979, para. 8.5). Concentration and centralisation of environmental regulation is therefore not necessarily desirable, especially if the aim is to integrate environmental considerations into economic decision making. Indeed, if environmental regulation is conceived not simply as a policing function but as an active agent for the promotion of sustainable practices, it is vital that regulatory bodies establish working relationships with all those who control the management and exploitation of natural resources. This is, indeed, a characteristic feature of the various conservation agencies in Britain (for example, the Countryside Commission, English Nature and the Forestry Commission) which combine regulatory and management functions.

Any practical proposals for reorganisation and amalgamation of regulatory functions must also be fully aware of the different regulatory cultures involved (Jordan, 1993). On a number of key dimensions, for example, the NRA and HMIP present strikingly opposite characteristics and orientations. Thus NRA inspectors typically are biologists, whereas HMIPs are chemical engineers. The one is oriented to maintaining ambient standards, the other to emission controls. The HMIP tends to pursue a self-trust basis of regulation whereas the NRA pursues more of a policing approach. The former emphasises a procedural rationality of checking that things are in order (referred to disparagingly as a 'tick box' approach); the latter emphasises the detective work involved in investigating the source of a pollution incident and the legalistic requirements of taking formal samples. Finally, NRA staff have a strong identification with a particular geographical area (one or more river catchments) whereas HMIP identify with particular firms or sectors, and these differences are reflected in the distinctive organisational structure of the two agencies.

Underlying much of the debate about the agency is an assumption that coordination is 'a good thing': but is it? One current view in administrative theory suggests that, far from seeking to eliminate duplication and overlap, effective coordination requires it: 'administrative performance suffers more from the over-use of specific controls than from a supposed overcontrol in terms of multiple and redundant controls' (Rhodes, 1991: 530). In other words, as it is probably impossible to create the perfectly coordinated control system, it is better to build a redundancy of controls into an administrative system. Redundancy enables the use of a wider selection of control devices and it allows a broader range of stakeholders and interests to participate in or be represented in the regulatory system. Thus an approach to reform that was sensitive to the redundancy argument might seek to incorporate and sustain both HMIP and NRA regulatory cultures within a new agency. Alternatively, rather than sacrifice the many benefits of the extant devolved and decentralised administrative system, the very notion of coordination within a single agency might be rejected.

Devolved, cooperative, tailor-made regulation is, in fact, the British tradition. It has manifest weaknesses; but, its supporters contend, it is better to tackle these, or at least ponder long and hard before abandoning long established administrative practices completely. Certainly, the more devolved regulation is, the more pressing is the need for detached strategic policy making and for open and accountable auditing of regulatory practice, both to ensure consistency between sectors and to guard against the 'capture' of regulators by the interests being regulated. That calls for a coherent framework for environmental protection, not necessarily a single grand authority.

Modern environmentalist thinking is firmly committed to the view that one of the major problems to be overcome is the sectoral perspective on the environment that permeates governmental policy and administrative structures in all countries. Instead, it is argued that governments need to adopt a more holistic strategy which recognises the interlocking nature of all environmental problems and integrates environmental policy across individual policy sectors. But, for the foreseeable future, the dilemma facing any government is how – in a sectorally-administered world consisting of different policy arenas, departments and disciplines – to control and implement cross-sectoral functions without compromising its strategic perspective. In Britain, this dilemma has been recognised (implicitly) since the rise of modern environmentalism in the late 1960s/early 1970s in the way that the Government has adopted two separate but parallel strategies.

First, it has tried to improve administrative coordination wedded to strong departmental authority (a strategy manifested in the establishment of the DoE in 1970). However, it is important to note that although the DoE was formed as a high level super-department combining a number of responsibilities and possessing Cabinet status, it was very much a ministry *of* the environment (of which the environment was but one (minor) responsibility alongside local government and housing), rather than a ministry *for* the environment. Gradual change in this direction is detectable, particularly with the declining importance of the housing portfolio and the growing political visibility of the environment. Unfortunately, the establishment of an independent agency might impede progress in this direction. For, in the past, where environmental functions have been allocated to various individual quangos (e.g., the National Parks, or the NCC, now known as English Nature) these bodies have subsequently been treated in Whitehall as sectional concerns seeking to influence government, rather than as central actors within the political process. The integrated agency may suffer a similar fate. Yet it is vitally important that the DoE be encouraged to provide the environment with a powerful, firmly rooted voice at the heart of central government. Care is needed, however, not to reproduce the situation found in some countries where a small specific Ministry of the Environment has been formed which possesses a clear sense of purpose but which finds itself isolated from the main policy making arenas and lacking muscle in inter-departmental power struggles.

The second strategy of the government has been to nurture and protect the holistic vision: manifested in the Royal Commission on Environmental Pollution which possesses no executive functions at all. The Commission has a remit to roam where it pleases through government with the purpose of following a scientific strategy directed by a broad, holistic aim. It is probably only because the Commission lacks regulatory powers that civil servants have been prepared to accept its free-ranging role. It is not surprising that it was the Commission that first proposed IPC in the 1970s: a concept that it failed to locate in any political context.

Thus this parallel strategy reflects a central dilemma in British environmental policy. On the one hand, a scientific strategy unencumbered by political or administrative responsibilities; on the other hand, a strategy of administrative coordination which often lacks a clear sense of environmental purpose. Viewed in this way, the formation of a cross-sector Environment Agency is simply the final (or latest) twist.

The specific decision to include the NRA replete with all its regulatory and managerial functions within this proposed agency against the preferences of business interests and the Select Committee for a smaller, HMIP-dominated institution suggests that there is also some instability in the established policy communities and networks that shape environmental policy. To understand this particular decision it is necessary to return to events prior to the establishment of the NRA. Policy making in the water industry sector, as illustrated by the re-organisation of 1973–4, had traditionally been characterised by technical considerations which reflected the dominance of engineers in the consensual decision-making process. However, by the mid-1980s further restructuring and strict financial restrictions had resulted in the ascendancy of economic over technical concerns. The privatisation of the water industry provoked enormous political controversy stretching far beyond the parochial world of the water policy community. For a brief period this policy community disintegrated (Richardson, Maloney and Rüdig 1992). At the heart of this instability was the issue of regulation. The original government plan for the privatised water authorities to regulate their own activity generated opposition from a wide range of interests which included the Institute of Water & Environmental Management (IWEM), the IEEP, the Council for the Protection of Rural England (CPRE), the Country Landowners' Association and the CBI. The lobbying of these groups contributed to the eventual decision to set up the NRA as the new public regulatory body with responsibility for the full water cycle and with a decentralised structure based on river basin boundaries. For a short period, a broader issue network was therefore able to influence government policy before consensus emerged and the closed policy community re-established. But the creation of the NRA – an unintended consequence of the privatisation programme – resulted in the emergence of a new issue network. The NRA is sponsored by the DoE, has policy links with MAFF and the Welsh Office, and an important relationship with the European Commission regarding EC environmental objectives. NRA responsibilities bring it into close contact with HMIP, Office of Water Services, agencies such as English Nature, local authorities and, obviously, the water industry. Beyond these public bodies there is a range of interest groups with a stake in the water environment and therefore affected by NRA activity: industrial and trade associations such as the CBI; farming interests, particularly the National Farmers' Union (NFU); riparian groups such as angling organisations; landowning interests such as the CLA; and many national and local environmental groups. The latter, in particular, were impressed by the NRA's commitment, expressed in its first annual report, to be an interventionist regulatory body and its subsequent readiness to prosecute polluting miscreants. But the NRA has also gained respect and support from many other members of this issue network. For example, the National Federation of Anglers, in its submission to the Environment Committee, declared that, 'in the short time since it was created, we have been impressed by the manner in which the NRA have carried out their duties' (House of Commons Environment Committee, 1991–2: 110).

Consequently, as we have seen, when the four options for the proposed agency emerged, many of the groups in this issue network lobbied against the dismemberment of the NRA. One of the NRA's most cogent defenders was the IWEM, which saw the NRA as the cornerstone of a new all-embracing agency and argued that: 'the strong and effective regional structure and the river basin boundaries of the NRA will provide a sound base for the new agency, in terms of both its water and waste regulation activities' (House of Commons Environment Committee, 1991–2: 84). Of course, the coalition of environmentalists, riparian, amenity, farming interests and professionals was opposed by other groups within the same issue network, notably business interest groups. Industrial interests were convinced that a smaller agency operating along the lines of the HMIP would be more sympathetic to them. Nevertheless the case for the defence of the NRA was made by a sufficiently broad and articulate lobby to ensure that this key decision on the future of pollution regulation would not be made entirely within a closed producer-dominated policy community. Indeed, at face value, it would appear that the 'outsiders' have won; the NRA will not be dismembered. It is, however, too early to say whether such an assumption is correct.

CONCLUSION

Since this chapter was written as the Environment Bill was passing through Parliament so it was unclear exactly what form the agency would take, and in particular how the contrasting regulatory cultures of the NRA and HMIP will be integrated, the following thoughts can only be speculative. It is widely assumed that by sheer weight of numbers – the NRA has some 8000 staff while the HMIP has just 300–400 – the NRA will 'swamp' the HMIP. Moreover, because of the circumstance of its origin and the breadth and variety of its issue network, the NRA is a more practised and successful lobbyist. In short, environmental groups are optimistic that the new agency will prove a good and effective 'friend' of the environment. But, as we have seen, the NRA issue network was not united on option three. Producer interests will be particularly reluctant to relinquish the benefits of the extant closed policy community that encloses the HMIP responsibilities. The fears of industry had been partially allayed by Howard's declaration that the agency will be seeking 'both effective protection to the environment and value for money'. Sure enough, the original Environment Bill, reflecting the Government's deregulation initiative, required the agency to take into account the costs and benefits of its proposed actions before deciding whether to proceed. It seems most unusual and inappropriate to give an agency certain duties and then to qualify them in this way by statute. A requirement to be aware of the economic implications of its actions would be better incorporated in ministerial guidance to the agency. Of course, if the requirement was to consider the full costs and benefits of actions and inactions (including the environmental costs) then there would be nothing to fear and a lot to be gained. But as it is, environmental groups are understandably wary of this whole matter, seeing it as a sop to polluting industries. There are fears, too, about the recent speculation that the NRA will soon declare its commitment to self-regulation. It would appear that industrial lobbying has not gone unnoticed. The victory of the NRA 'lobby' may yet prove to be hollow; whilst the NRA may not be dismembered it may be denuded of its ambitions.

The Bill does not give the agency clear aims and objectives but specifies that ministers will give it periodic guidance on how it should contribute to 'sustainable development'. The NRA has benefited from having clear objectives set out in statute. It seems then that the Bill would weaken the agency's commitment to be an advocate of environmental protection. Nor is there any suggestion that the agency will acquire extra powers on the lines recommended by O'Riordan and Weale (1992). On the contrary, the original draft of the Bill proposed to remove the duty, currently held by the NRA, to 'further' conservation interests. This led to a storm of protest from environmental groups to which John Gummer, Secretary of State for the Environment, partially responded by restoring the duty to 'further' conservation to all except the agency's pollution responsibilities. This is likely to be the key battleground between environmentalists and industrialists.

In effect, as Haigh (1993) has observed, the current plan is to create a pollution control agency rather than an environment agency. Within this narrower definition it is not yet clear whether the agency would be able to offer advice or make recommendations for government pollution control policy. Public policy theory makes clear that implementation should be seen as an essential part of the policy process. As the regulatory body responsible for implementing pollution control policy the agency will be in a good position to offer informed policy recommendations, but there is unlikely to be much support in government circles for it to play an active role in policy advice. Yet without such a role the agency may represent little more than 'old wine in new bottles'.

Note

A shorter version of this paper appeared in *Political Quarterly*, July–September 1994.

References

Carter, N. (1992), 'Whatever Happened to the Environment? The British General Election of 1992', *Environmental Politics*, 1, 442–8.
Department of the Environment (1988), *Integrated Pollution Control*, Consultation Paper, London.
Department of the Environment (1991), *Improving Environmental Quality: the Government's Proposals for a New Independent Environment Agency*, Consultation Paper, London.
Department of the Environment (1992), *The Government's Response to the Second Report from the House of Commons Select Committee on the Environment, Coastal Zone Protection and Planning* (London: HMSO), Cmnd 2011.
ENDS Report No 198.
Haigh, N. (1993), 'Who Protects the Environment?: An Environmental Agency in Perspective', paper for the *NSCA Environmental Protection Conference*, October.
House of Commons Environment Committee (1988–9), *Toxic Waste*, Session 1988–9, Second Report (London: HMSO).
House of Commons Environment Committee (1991–2) *The Government's Proposals for an Environment Agency*, Session 1991–2, First Report (London: HMSO).
Jordan, A. (1993), 'Integrated Pollution Control and the Evolving Style and Structure of Environmental Regulation in the UK', *Environmental Politics*, 2, 405–27.
National Audit Office (NAO) (1992) *Coastal Defences in England* (London: HMSO), Report No.9.
O'Riordan, T. and Weale, A. (1989), 'Administrative Reorganisation and Policy Change: The Case of Her Majesty's Inspectorate of Pollution', *Public Administration*, 67, 277–94.
O'Riordan, T. and Weale, A. (1992) *Greening the Machinery of Government* (London: Friends of the Earth).
Rhodes, R. (1991), 'Interorganisational Networks and Control', in F.-X. Kaufman (ed.), *The Public Sector: Challenge for Coordination and Learning* (New York: De Gruyter).
Richardson, J., Maloney, W. and Rüdig, W. (1992), 'The Dynamics of Policy Change: Lobbying and Water Privatisation', *Public Administration*, 70, 157–75.
Royal Commission 7th Report (1979).
This Common Inheritance: Britain's Environmental strategy (1990) (London: HMSO) Cmnd 1200.
Vogel, D. (1986) *National Styles of Regulation* (Ithaca, NY: Cornell University Press).

5 The Precautionary Principle in UK Environmental Law and Policy

Andrew Jordan and Timothy O'Riordan

INTRODUCTION

In the last decade the precautionary principle has entered into the lexicon of modern environmentalism with remarkable speed and stealth. Nowadays, it appears regularly in national legislation, in international statements of policy and in the texts of international conventions and protocols. More recently, it has been adopted as a guiding principle of environmental policy in both the EU and the UK, and it makes an appearance in the 1992 Rio Declaration (a statement of principles and general obligations to guide the international community towards actions that promote more environmentally sustainable forms of development). So frequently is the term invoked that Cameron and Abouchar (1991: 27) go as far as to posit that, if the present rate of proliferation is sustained into the future, precaution may well become '"*the*" fundamental principle of environmental protection policy and law at the international, regional, and local scales' (emphasis in original). Even the more cautious of the principle's advocates suggest that 'the concept has at least approached the status of a rule of customary international law' (Hey, 1992: 307).

Given such ambitious claims, it may seem somewhat strange to discover that the precautionary principle has neither a commonly agreed definition nor a set of criteria to guide its implementation. Precaution is a slippery but appealing term that emerged in the social democratic planning era of the 1970s in the former West Germany. At the core of the early conceptions was the belief that regulatory agencies and governments should move to minimise environmental risks by anticipating possible danger and if possible preventing it, but even this was surrounded by layers of ambiguity. Today there exists little consensus about what the precautionary principle really means; on how it might be implemented; on its implications for policy making, scientific research and decision making generally; and how it might be reconciled with other, well established, principles of modern environmental law and policy (e.g.,'the polluter

57

pays', prevention). The precautionary principle seems to be infinitely contestable, but then so is sustainable development, a concept with which the precautionary principle has a great deal of affinity. Underlying both are enormously complex trade-offs between human needs and environmental rights; landscape development and nature conservation; immediate consumption and long term well-being.

In the first part of this chapter we sketch out our own interpretation of the precautionary principle based on seven core concepts. We base our interpretation upon an historical overview of the principle's development and steady proliferation since the 1970s. In the second section, we consider the manner in which the principle has been interpreted by policy makers first in Germany and then in the UK, and then, upon this analysis, we examine the extent to which it finds expression in, or informs the development of, UK policies relating to the environment. In the final section we present four short case studies which highlight instances in which the UK Government has both resisted and promoted the application of precautionary measures in the more recent past. The conclusion from this section is that while the principle may be implicit in much UK legal language, in practice it is still interpreted in favour of economic development rather than nature conservation.

THE PRECAUTIONARY PRINCIPLE IN CONTEMPORARY ENVIRONMENTAL LAW AND POLITICS

In many respects, the precautionary principle can be likened to the concept of sustainability for it is at once both vague and complex; broadly appealing but a cause for suspicion. 'There is', as Freestone (1991: 30) posits, 'a certain paradox in the widespread and rapid adoption of the precautionary principle'. While it is universally and increasingly applauded as a 'good thing', no one is quite sure what it really means, or how it might be implemented. In fact, precaution, like sustainability, acts as a repository for a jumble of adventurous beliefs and challenging concepts. Both take their cue from changing social conceptions about the roles of science, economics, ethics, politics and the law in proactive environmental law and management (O'Riordan and Cameron, 1994). To those familiar with environmental management issues, however, 'sustainability' and 'precaution' will appear as a new gloss on much older, but as yet unresolved, conflicts and dilemmas.

The precautionary principle is framed in culture and politics and its importance lies in the way in which it captures misgivings about the scientific method, about the proper relationship between human needs and the health of the natural world, and about the appropriate way to incorporate the rights of future generations into the policy making process (O'Riordan, 1993). The Oxford Concise Dictionary defines precaution as 'an action taken beforehand to avoid risk or to ensure a good result'. The central role of the precautionary principle is to guide administrators and regulators who are required to make decisions or develop policy in circumstances where scientific information is imperfect and the likely impact of their actions is uncertain (Wynne, 1992). The precautionary principle demands that action be taken to prevent environmental damage even if there is uncertainty regarding its cause and possible extent, on the grounds that it is better to be roughly right in due time, bearing in mind the consequences of being very wrong, than to be precisely right too late. In other words, the environment should not be left to show harm before protective action is taken; scientific uncertainty should not be used as a justification to delay measures which protect the environment. The purpose of the precautionary principle is 'to encourage – perhaps even oblige – decision makers to consider the likely harmful effects of their activities on the environment before they pursue those activities' (Cameron and Abouchar, 1991: 2). For Weale (1993: 209), the broader implication is that politicians and policy makers should act with 'due care and attention in the face of scientific uncertainties'. For example, regulators might wish to instruct a proto-polluter to reduce or terminate emissions at source, even in the absence of sufficient evidence to prove a causal link between emissions and adverse environmental impacts, and prior to a wider consideration of risks and the costs and benefits to society as a whole. Decision makers can also exercise precaution by preserving, rather than developing, natural assets or ecological processes that are judged to be of special value, such as wetlands or ancient woodlands. This is currently the case with respect to the ozone layer: for better or worse, successive declarations regarding the removal of ozone depleting substances are now aimed at protecting that part of stratospheric ozone that remains unmolested (Rowlands, 1993).

Those more sceptical of the precautionary principle are more likely to assert that it is only popular because it is so comfortingly vague; that it fails to bind anyone to anything or resolve some of the deeper dilemmas which characterise modern environmental policy making (Weinberg, 1985; Bennett, 1992). There are legal scholars, for example, who consider precaution to be too blunt an instrument to act as a regulatory standard or principle of law. Bodansky (1991: 5), for example, is highly suspicious of the precautionary principle because it 'does not specify how much caution should be taken' in any situation. It cannot, for example, determine what is an acceptable margin of error or what exact threshold of risk warrants the application of precautionary actions. Nor can it determine when precautionary measures should be taken, or define the point at which abatement costs become socially or environmentally 'excessive' (Bodansky, 1991: 5). In a review of German policy, Boehmer-Christiansen (1994) concludes that 'the precautionary principle has little meaning other than that of enabling the policy process to attempt environmentally more ambitious solutions'; Bodansky (1991: 5) suggests that it constitutes little more than 'a general approach to environmental issues'.

The emergence of the principle has also engendered a very lively debate within the scientific community. Gray (1990)' for example, denies the principle any role in scientific research, 'since by definition it does not have to rely on scientific evidence!'. He believes that it is, at best, an 'environmental philosophy'; purely a matter for administrators and lawyers rather than scientists seeking 'objective scientific evidence'. This view is, however, contested (Johnston and Simmonds, 1990; Wynne and Meyer, 1993).

It would also be misleading to portray precaution as an entirely novel concept, for precaution-based measures are already institutionalised in UK law and regulatory practice. All too often, however, such provisions are weak and discretionary, for they are biased in favour of development. The Alkali Act (1906) laid down fixed statutory emission standards for certain types of process rather than the normal requirement to use the 'best practicable means' (McLoughlin and Forster, 1981: 71). The EPA introduced a prior authorisation procedure for operators scheduled for integrated pollution control (Jordan, 1993) which prohibits the use of certain processes until they are fitted with the best available abatement technology. Operators are required to ensure that certain emissions are rendered 'harmless and inoffensive' to the environment, but this is qualified with the powerful proviso that 'excessive costs' should not be incurred. 'Prior authorisation' and the use of 'best technology' are keystone principles of a system of air pollution control which was inaugurated with the passage of the first Alkali Act in 1874. Meanwhile, Part VI of the 1990 Act (which relates to the release of genetically modified organisms, or GMOs) is probably the most precautionary piece of legislation ever introduced in the UK. The Act provides for strong, source based controls over the release of GMOs. Here, regulation will err on the side of caution while the commercialisation of biotechnology is still in its infancy (Tait and Levidow, 1992). In principle, the application of environmental impact assessment (EIA) is designed to predict possible adverse ecological and social outcomes arising from certain development proposals, and then to make recommendations as to how they might be prevented or mitigated.

All these examples indicate how an element of precaution has always played some role in UK environmental law and policy. However, three trends in contemporary environmental politics are helping to produce a more propitious context in which the precautionary principle can flourish.

First, the science of assimilative capacity, predictive modelling and compensatory investment to offset the loss of ecological resilience is being challenged. This challenge takes a number of forms, including a critique of the role of authoritative science in policy formulation (Barker, Peters and Guy, 1993), a reassessment of scientific information in risk regulation (Jasanoff, 1990), and a more populist argument that 'conventional science' is no longer the only route to judgements about ecological futures (Wynne and Meyer, 1993). The outcome of this is a feeling that, at the very least, science should evolve into a more applied interdisciplinary format for coping with environmental threats, and that it should be seen as a tool for a more open and participatory culture of decision taking.

The second is the ever growing scale and scope of environmental politics and policy making (McGrew, 1993: 23). The freedom for manoeuvre for the nation state to control its environmental policy has been reduced, certainly in Europe, since the mid-1970s, and will accelerate in the prolonged aftermath of the 1992 Rio Summit. As the locus of environmental action moves beyond national borders, individual countries will find that they are forced to adopt tougher norms and standards that are setting the lead in other parts of the world. The rapid spread of the precautionary principle from international to domestic environmental legislation bears testimony to this trend. In turn, the application of precautionary measures in domestic contexts is driven, in large part, by events on the international stage. The interaction is dialectical. For example, political obligation (underpinned by the legal concept of 'common but differentiated responsibility': Porras, 1993) and the pressure from 'leader' states, will impel 'laggard' states to do more than they might have done in the absence of international regulatory regimes (Keohane, Haas and Levy, 1993: 16). Stricter environmental standards implemented by the UK Government as a result of tightening international norms, and then justified on the basis of the precautionary principle, are discussed more thoroughly below.

Third, the stringency with which the precautionary principle is applied depends upon and is also a useful barometer of deeper social and economic changes. Precautionary measures, for example, are more likely to be applied when public opinion is instinctively or knowledgeably risk averse. This is certainly the case with GMOs and the most toxic and bioaccumulative substances. The application of precautionary measures is welcomed and supported in societies where there is high concern for environmental quality and other 'post-material' concerns. Dunlap, Gallup and Gallup (1993) intimate that risk aversion over vulnerable natural systems is by no means confined to groupings of middle class liberals in wealthy societies. Their research supports the view that precaution is becoming an established human outlook in all walks of life and countries. Thus, the rise of precautionary thinking cannot be divorced from much deeper transformations that have occurred in social attitudes over the last 20 years or so: to wit, a growing concern for post-material ('higher order') values (Inglehart, 1989). These social trends support and give potency to the precautionary principle, or at least a more precautionary environmental ethic, even when it lacks a specific legal definition and a concrete meaning. In contrast, the precautionary principle is less likely to evolve in regulatory regimes where environmental considerations do not yet figure prominently, even when there is public support for the broad principle. This appears to be the case in Eastern and Central Europe, where the exigencies of simply keeping alive and establishing a foothold into the market economy force people to place more emphasis on issues of immediate survival rather than long term welfare and environmental well-being. It is also notable that the precautionary principle first flourished in Germany – the largest and richest country in the EU – at a time when social democratic planning was in vogue and the economy was growing. In the early 1990s, and with its economy lodged in a deep recession, the UK Government seems more pre-occupied with de-regulation and 'lifting' the burden on industry than securing higher levels of environmental protection (*The Economist*, 1993). In a similar vein, precaution is less welcome in right of centre, free market supporting cultures (e.g., in the USA and Japan), and in hierarchically structured professions, such as accountancy and international finance.

CORE ELEMENTS OF THE PRECAUTIONARY PRINCIPLE

From the complex, and at times confusing, debate on the meaning and applicability of the precautionary principle it is possible to abstract a number of commonly occurring themes. We have distilled these down to seven interlinked concepts which we believe lie at the nucleus of present day thinking on precaution (O'Riordan, 1993: 3). These are: a willingness to take action in advance of formal justification of proof; proportionality of response; a preparedness to provide ecological space and margins for error; a recognition of the well-being interests of non-human entities; a shift in the onus of proof on to those who propose change; a concern with future generations and dysgenic impacts; and a need to address ecological 'debts'. By no means are all these dimensions formally approved of in existing law and common practice. Indeed, the precautionary principle will always be slightly protean because it is evolving through distinct pathways in different countries and at different rates.

Proaction: this concerns a willingness to take action in advance of scientific proof of evidence on the grounds that further delay may prove to be ultimately more costly to society and nature. Precaution is not simply the prevention of manifest or predicted risks that have been scientifically proven (Cameron and Abouchar, 1991: 9). Rather, the precautionary principle goes beyond the notion of prevention in the sense that it insists that policy makers move to anticipate problems before they arise or before scientific proof of harm is established (Freestone, 1991). In practice, the line differentiating precautionary measures from preventative actions may be very blurred. A very strong formulation of precaution demands that all risks to the environment are minimised, that emissions are reduced as low as possible, and that important (or 'keystone') ecological assets and processes are maintained intact. Greenpeace, for example, go as far as to say that 'the precautionary principle calls for the prohibition of the release of substances which might cause harm to the environment even if insufficient or inadequate proof exists regarding the causal link' (emphasis added: Horsman, 1992: 76). This 'strong' conception offers scientific judgements a very limited role in decision making.

In reality, however, precaution is often linked to some consideration of risks, financial costs and benefits. In Germany, for example, the application of precaution is tempered by two other general principles of law: proportionality of administrative action and avoidance of excessive costs (Von Moltke, 1988: 68). Although the principle of precaution does not state how these different factors should be traded off, it still suggests that a strenuous search be conducted for alternative modes of development which minimise discharges and waste products, regardless of whether they are known to have harmful effects, on the basis that prevention is often, though not always, more cost effective than cure. Moreover, the closer controls are placed to the source of the emission and the earlier environmental considerations are factored in to decision making, the more precautionary the overall trajectory of development will be.

Cost-effectiveness of action, or proportionality of response, is important in showing that there should be a regular examination of the identifiable social and environmental gains arising from a course of action that justifies the costs. It is clear that the adoption of a very strict interpretation of the precautionary principle can be costly, especially if the benefits forgone are appreciable and new information reveals that the precautionary measure was in fact unwarranted. It is possible to differentiate between actions taken on a precautionary basis (on the strictest interpretation this would equate with zero emissions) and the 'optimal' level of pollution that is more likely to be elucidated by a fuller risk-benefit or cost-benefit analysis (Pearce, Turner and O'Riordan, 1993). The difference between the two would represent the social opportunity cost of adopting a precautionary perspective. In some instances the net social cost of adopting a precautionary perspective could be very large, especially if the adverse environmental impacts turn out to be less important than predicted. However, cost-benefit analyses presuppose good scientific understanding and sufficient time to undertake them but these factors cannot always be relied upon.

Clearly, there is no easy way of integrating risk, financial considerations and highly uncertain science into decision making with the appropriate degree of timeliness. Neither does the precautionary principle provide a rigorous mechanism for balancing these disparate factors. However, it does insist that standard techniques of cost-benefit analysis should incorporate the wider social and environmental costs of development, and that it should be biased in favour of high environmental quality (e.g., development should only be allowed when the benefits of doing so are *much* greater than (rather than simply greater than or equal to) the associated costs: Pearce, 1994).[1]

Safeguarding ecological space: the precautionary principle insists that the overall capacity of environmental systems which acts as a buffer for human well-being must be adequately protected: 'any error in risk calculation should be to the advantage of the environment' (Bodansky, 1991: 5). This entails leaving a sufficiently wide natural cushion in the functional equilibria of natural systems. In effect, this means that humans must learn to widen the assimilative capacity of natural systems by deliberately holding back from unnecessary and environmentally unsustainable resource use on the grounds that exploitation may prove to be counter-productive, excessively costly or simply unfair to future generations. Nature's assimilative capacity cannot always be taken for granted: often, we simply do not know enough about the chaotic behaviour of natural systems (Gleick, 1987; Lorenz, 1993), to be able to identify 'critical thresholds'[2] with any degree of certainty (Stebbing, 1992; McGarvin, 1994).

In terms of policy, the precautionary principle offers an explicit challenge to the so-called 'dilute and disperse paradigm' (Group of Experts in the Scientific Aspects of Marine Pollution–GESAMP, 1986) which so often informs the practice of setting regulatory standards in the UK. The paradigm is built upon the assumption that science can determine the ability of an ecosystem to assimilate hazardous substances without incurring long term damage, and that individual emission permits can be tailored consonant with these 'safe' margins. The precautionary 'paradigm' is based on the denial of the general validity of these assumptions (Hey, 1992: 308): that is, that science does not always provide the insights with the necessary degree of timeliness or accuracy, and that preventing emissions is often, though not always, a better course of action than restoration of damage once it has occurred.

Environmental well-being has intrinsic value and legitimate status: precaution goes to the heart of the philosophical and political debate on the proper relationship between humans and the non-human ('natural') world (Dobson, 1990; Eckersley, 1992). In promoting a more humble and less rapacious attitude to the environment, the precautionary principle presents a profound challenge to some of the unstated assumptions of 'modern' (and particularly Western) societies: material growth, the power and efficacy of scientific reason and the pre-eminence of human interests over those of other entities. Boulding (1966) used the term 'cowboy' to characterise the human race's perception of the natural environment as an 'open' resource and an infinite sink; a limitless domain to be conquered and exploited. The human race is a colonising species without an institutional or intellectual capacity for equilibrium, and notions of long term 'care', 'precaution' and 'restraint' strike at the very heart of its common purpose (Jordan and O'Riordan, 1993). In contrast, the precautionary principle lends strength to the notion that natural systems have intrinsic rights and a non-instrumental value that should be accounted for in decision making.

Unfortunately, the precautionary principle does not state how much environmental quality should be sacrificed for material growth, and neither does it determine how a 'non-instrumental' respect for nature should be incorporated into decision making. However, it does offer a strong presumption in favour of high environmental protection and a justification for treating certain environmental functions or features as inviolable. This is a prospect that usually causes alarm amongst those who believe that such a concept is an excuse for deep ecology to ride roughshod over 'sensible' forms of development or impose 'limits' to material growth. The US lawyer, Christopher Stone (1987), has sought to allay these fears by indicating that a creative partnership in law can be established to allow nature rights of existence that are not absolute, but require careful deliberation before being set aside.

Shifting the onus of proof: most interpretations of the precautionary principle suggest that the burden of proof should, by some degree, be shifted on to the proto-developer of the environment and away from the proto-victim

(Cameron and Wade-Gery, 1992). Traditionally, the law has tended to privilege parties accused of degrading the environment rather than the victims of pollution (i.e., those that are forced to bear the external costs). In general, claims for damage in the law are only upheld if the victim can prove that the emission (or the damage) was reasonably 'foreseeable'. 'Acts of God' and 'accidents' tend to disallow claims for compensation. In this sense, the law offers little inducement to developers or operators of industrial processes to take adequate precautions with regard to the environmental impact of their actions. The introduction of a *strict liability* regime, on the other hand, would only require the victim to prove that the polluter failed to act with 'due diligence' to gain compensation; in the case of absolute liability, the victim would merely need to prove that damage had occurred to gain financial restitution. More stringent still, would be to reverse the burden of proof entirely (i.e., the burden of proof is placed upon the proto-polluter to prove that emissions are 'harmless' before the activity is sanctioned), as in the licensing of new medicines. But this presupposes some measurement of 'harmlessness', which is sometimes as equally as difficult to prove as 'harmfulness'. Reversing the burden of proof entirely would be radical, since it would impose an explicit and legally binding 'duty of care' on those who propose to alter the status quo. Admittedly, this raises profound questions over the degree of freedom to take calculated risks, to innovate, and to compensate for possible losses by building in ameliorative measures. Yet UK law, in a sense, is already moving in this direction with the introduction of formal duties of care, set against the backdrop of a broader, but intensifying, debate on the possible extension of strict liability for environmental damage, no matter how anticipated.

Futurity issues: the application of precaution extends the scope of environmental policy from certain and known problems that occur in the present, to future and more uncertain issues. Precaution urges politicians to act with due care and diligence: to anticipate and take action against problems before they occur. Precautionary actions could be considered as an investment (or insurance) against unforeseen mishaps, or the acceptance of higher costs now in order to guard against dysgenic impacts. But it also implies committing current resources to investments for the future, the benefits of which may be uncertain or, at worst, non-existent. Since conclusive scientific evidence of harm or excessive damage in the future may not always be available to justify the commitment of resources to precautionary investments, other grounds for legitimation may need to be present: namely moral, political, ethical and legal (O'Riordan, 1992: 3).

Paying for ecological debts: precaution is essentially forward looking but there are those who recognise that in the application of care, burden sharing,

:cologically buffered cost-effectiveness and shifting the burden of proof
here ought to be a penalty for not being cautious or caring in the past. In
>ther words, the scope of environmental policy should also be extended
>ackwards in time. In a sense this is precaution in reverse: compensating for
>ast errors of judgement based on ignorance or an unwillingness to shoulder
ın unclearly stated sense of responsibility for the future. This element of
>recaution is still embryonic in law and practice, but the notion of 'common
>ut differentiated responsibility' enshrined in the Climate Change
Convention and the concept of conducting precaution 'according to capabilities' as noted in Principle 15 of the Rio Declaration, reflect these ideas to
some extent. 'Precaution in reverse' is implemented by the retrospective
application of duty of care and strict/absolute liability regimes.

In each of the seven dimensions it is possible to envisage precaution
acting along a continuum. The weaker (less stringent) formulations, for
example, are still protective of the legal status quo on burden of proof and
risk avoidance and tend to be restricted to the most harmful and bioaccumulative substances and situations where costly and irreversible damage
appears likely. They advocate a role for biased cost-benefit analysis, incorporate a concern for the technical feasibility of measures, and tend not to
seek to go beyond scientific proof of harm. Much stronger formulations
prescribe very low or even zero emission limits, a drive to introduce the
best technology regardless of cost, a reversal of the normal burden of proof
and the absolute protection of critical natural habitats. At present, the
common line appears to lie somewhere between these extremes, namely to
act prudently when there is sufficient scientific evidence and where action
can be justified on reasonable judgements of cost-effectiveness and where
action could lead to potentially irreversible harm. In practice, the application is usually derived for chemicals whose effects are potentially toxic,
persistent and bio-accumulative (i.e., only the most dangerous).[3] More
recently however, the precautionary principle has been extended to cover a
wider array of less toxic substances, greenhouse gases and even sustainable
development.[4] It is important to note that each element also requires an
enabling mechanism, which could be legal (e.g., strict liability), economic
(e.g., a safe minimum standards) or technological (e.g., clean production).

PRECAUTION AS AN EVOLVING PRINCIPLE OF INTER-
NATIONAL ENVIRONMENTAL LAW AND POLICY

Those who have sought to understand the precautionary principle have
traced its evolution back to the concept of *Vorsorge*, which developed in

the former West Germany during the 1970s. *Vorsorge* can be interpreted
as 'foresight' or 'precaution', although in German usage it also implies
notion of good husbandry and 'best practice' (Rehbinder, 1988). It is by
no means easy to explain what the *Vorsorgeprinzip* ('precautionary princi
ple') means, for it must be considered in the context of Germany's institu
tional structure and legal traditions. *Vorsorge* made an early appearance in
the federal government's 1976 environmental report: 'Environmenta
policy is not fully accomplished by warding off imminent hazards and the
elimination of damage which has occurred. Precautionary environmenta
policy requires furthermore that natural resources are protected and
demands on them are made with care'. In general, *Vorsorge* comes into
play when the risks of environmental damage 'are not (yet) identifiable, or
even in the absence of risk' (Von Moltke, 1988: 61). In principle
Vorsorge implies that authorities should move to minimise all risks, but in
practice *Vorsorge* tends to be married to the concept of 'proportionality'
(which encompasses issues of economic cost, technical and administrative
feasibility): (Von Moltke, 1988: 68). Vorsorge is implemented by way of a
requirement, placed on operators of scheduled processes (e.g., power sta
tions), to use the very best technology (or *Stand der Technik*: Boehmer
Christiansen and Skea, 1991). It is notable that the 'technology forcing'
element of *Vorsorge* has promoted the development of a clean technology
industry which in turn has delivered enormous social and economic
benefits.[5] For the Germans, then, the application of precaution is capable
of stimulating environmentally sustainable growth.

In practice, *Vorsorge* is less a well honed principle of law and more a
philosophical presupposition in favour of more stringent environmental
protection measures. Throughout the 1980s – a period when the German
economy experienced strong growth and the Green political parties cap
tured an increased share of the popular vote – *Vorsorge* acted as an import
ant animating concept in German environmental policy. It was used to
justify the imposition of very rigorous protection measures across a range
of environmental policy issues such as acid rain, global warming and pol
lution of the North Sea (Von Moltke, 1991; Weale, 1992: 66–092). In con
trast, the British Government's response across all three issue areas was,
initially at least, to procrastinate on the grounds of 'excessive cost' and
scientific uncertainty (Boehmer-Christiansen, 1992; Brown and Jordan,
1993; Maxwell and Weiner, 1993).

For the Germans, then, precaution is an interventionist measure, a
justification for the state's involvement in the day to day lives of its
Länder and its citizenry in the name of wise environmental practice. It is

unlikely that other countries will choose to implement precaution in quite the same manner as the Germans.

Since then, the precautionary principle seems to have captured the imagination of lawyers, scientists and policy makers the world over. Indeed, the precautionary principle could be said to have taken on a life of its very own, since it is now to be found in almost all recent international 'hard law' treaties and conventions, as well as Ministerial declarations, and other examples of international 'soft law' (see Cameron and Wade-Gery (1992) for a thorough review). In each case, the exact order and form of words varies; different interpretations give slightly different emphasis to considerations of risk, economic cost and technical feasibility. For example, the 1990 Ministerial Declaration on the North Sea should be considered as a relatively strong formulation of the precautionary principle, for it requires states to adopt the best available techniques to reduce certain emissions 'even when there is no scientific evidence to prove a causal link between emissions and effects (the "principle of precautionary action")'. The fifteenth principle of the 1992 Rio Declaration, on the other hand, adds much greater weight to the 'proportionality' concept: 'In order to protect the environment, the precautionary approach shall be widely applied by States according to their capabilities. Where there are threats of serious or irreversible damage, lack of full scientific certainty shall not be used as a reason for postponing cost-effective measures to prevent environmental degradation'. A close reading of this statement, reveals that, *inter alia*, lower income countries may be permitted to apply precaution in a less stringent manner than higher income states; that it applies, first and foremost, to circumstances where damage is acute and probably human life-threatening, rather than general instances of damage; that economic efficiency considerations should also be taken in to account; and that actions to protect the environment should ultimately be cost-effective. It should also be apparent that, acting in concert, these factors strengthen the 'proportionality' element, which in turn counter-balances the proactive thrust of some of the stronger formulations of the principle.

The precautionary principle has also left its mark in European regulatory practice and with the full ratification of the (Maastricht) Treaty on European Union, precaution becomes a guiding principle of the EU's environmental policy. The Treaty notes that:

Community policy on the environment shall aim at a high level of protection taking into account the diversity of situations in the various regions of the Community. It shall be based on the precautionary principle and on the principles that preventive action should be taken, that environmental damage should, as a priority, be rectified and that the polluter shall pay.

This interpretation also infers that the precautionary principle has an explicitly spatial dimension: that is, what is judged to be precautionary in one locale may not be precautionary in another. But again, no indication is given as to how this might be implemented across a Community of member states, each with its own institutional traditions, geographical features and political complexions. Other than the above, precaution has not been defined by the European Commission, nor yet has the European Court been asked to provide an interpretation. Until a more thorough interpretation is provided the exact status of the principle remains uncertain. In general, and as Haigh (1994) rightly notes, precaution should apply when the Community moves to formulate future policies (i.e., it will not apply retrospectively), and precaution will not apply to those parts of individual member states' domestic policies which do not fall within the aegis of EU policy. In practice, however, most of the member states have already made declarations in other fora which express support for the precautionary principle.

Although precaution has been adopted as a general principle of EU policy, uncertainty remains over how it will be implemented and in what time frame. In a very useful assessment of the legal and policy implications of the Maastricht Treaty, Wilkinson (1992) offers the following, somewhat cautious, judgement:

[h]ow the principle should be applied in principle is by no means clear, but it might, for example, include the requirement that protective measures should be developed before specific environmental hazards are evident, and that the onus of proof (of there being no damage) should be placed on the polluter ... [W]ith the precautionary principle now informing the EC's environmental policy, pressure can legitimately be applied for the application of tighter pollution control standards. (Wilkinson, 1992: 224).

THE EMERGENCE OF THE PRECAUTIONARY PRINCIPLE IN UK ENVIRONMENTAL POLICY

In 1990, the UK Government enunciated five principles to guide future policies regarding the environment. One of these was precaution (HM. Govt, 1990: 10).[6] The Government's 1994 *Strategy on Sustainable Development* (HM Government, 1994: 34) also includes precaution as one of a number of 'basic aims and principles supporting sustainable development' in the UK (HM Government,1994: 7). Clearly, there are many and varied links between the concepts of precaution and sustainability, but unfortunately the government failed to specify or discuss them.

The very fact that government documents and political pronouncements relating to the environment now regularly refer to the precautionary principle, is of considerable interest in itself. First it attests to the fact that environmental policy is becoming more and more an international activity that transcends state boundaries, channelled by 'epistemic' networks (Haas, 1990) and transnational policy communities. The extension of *Vorsorge* from Germany to the level of international treaty making, then back down to the UK domestic scene is very much a symptom of the Europeanisation of environmental policy (Liefferink, Lowe and Mol, 1993). Second, precaution is now one of a number of principles which the UK Government identifies as the foundations for its environmental policy (HM Government, 1990: 10). To start with basic principles and arrive at an overall policy is a practice that is extant in other European states and, to a lesser degree, the EU; but it is foreign to the UK (Weale, 1992: 82). It remains to be seen what influence, if any, these principles might have in the development of policy, or the actions of ministers, statutory agencies and pressure groups. Finally, some of the more stringent formulations of the principle sit uncomfortably with the traditionally 'British' style of environmental policy making, which has been to emphasise the importance of balancing cost, risk and benefit factors in order to elucidate the most efficient use of the environment's innate capacity to absorb waste (Royal Commission on Environmental Pollution, 1984: 40). The British 'approach' is characteristically pragmatic rather than radical; tactical rather than strategic; reactive rather than proactive (Lowe and Flynn, 1989: 256). On many occasions, the UK Government has sought to delay the imposition of more stringent protection measures on the grounds of 'excessive cost' (Lean, 1991: 23) or 'uncertain science' (Boehmer-Christiansen, 1988), only to fall in to line under pressure from other states. Precaution, on the other hand, seeks to curtail the ability of politicians to invoke scientific uncertainty as a justification for avoiding or delaying the opposition of more stringent protection measures.

In fact, the imprint of the British 'approach' is discernible in the formulations of the precautionary principle which the UK Government has chosen to adopt. Most of these accept that precautionary actions may be necessary in situations where there are significant risks of irreversible harm, but this is also alloyed with a strong emphasis on the balancing notion of 'proportionality' of action (HM Government, 1994: 7). The following interpretation, which is to be found in the 1990 White Paper, is probably the most detailed and authoritative to date:

Where there are significant risks of damage to the environment, the Government will be prepared to take precautionary action to limit the use of potentially dangerous materials or the spread of potentially dangerous pollutants, even where scientific knowledge is not conclusive, if the balance of likely costs and benefits justifies it. The precautionary principle applies particularly where there are good grounds for judging either that action taken promptly at comparatively low cost may avoid more costly damage later, or that irreversible effects may follow if action is delayed. (HM Government, 1990: 11).

In the 1994 *Strategy*, the government claimed that this interpretation 'is consistent with others in international usage' (HM Government, 1994: 33), although that claim should be treated with caution. In the 1990 White Paper, it also added the following, which captures one of the central dilemmas facing all governments, and one implicit in both the principle of precaution and the concept of sustainability: 'Just as we believe that it is irresponsible for Government to be extravagant with taxpayers' money, so we see even stronger arguments against wasting the world's, or this country's, natural resources and bequeathing a burden of environmental debts tomorrow' (HM Government, 1990: 11). Meanwhile, in the DoE's 1993 consultation paper for the strategy, precaution appears as follows:

Decisions about levels of [environmental] protection should be based on the best possible economic and scientific analysis... Where appropriate (for example, where there is uncertainty combined with the possibility of irreversible loss of valued resources), actions should be based on the so-called 'precautionary principle' if the balance of likely costs and benefits justifies it. Even then the action taken should be in proportion to the risk. (DoE, 1993: 10)

However, in a subsection of the 1990 White Paper entitled 'Fact not fantasy: best evidence', the government made clear that environmental decisions need to 'look at all the facts and likely consequences of actions on the basis of the best scientific evidence available. *Precipitate action on the basis of inadequate evidence is the wrong response*' (emphasis added: HM Government, 1990: 11).

A close reading of these statements and the reports from which they are drawn, reveals much about the UK Government's attitude to precaution, and some of the factors which might have informed it. First, precaution is seen to be applicable in a wide variety of contexts, although exactly how it might be implemented, or in what time frame, is not revealed. The government states, rather vaguely, that 'the principle can be applicable to all forms of environmental damage; nor should it apply only to the actions of government' (HM Government, 1994: 33). From this, it is not clear whether, for example, the taxpayer or the polluter should bear the costs of precautionary actions, or how precaution might be incorporated in to the process of setting (say) emission standards or water quality objectives. Second, emphasis continues to be placed on the value of 'sound science': namely that, at some point in the future, scientific evidence of sufficient authority will be forthcoming to vindicate the need for a particular course of action. In doing this, the government has chosen to adopt a much weaker formulation of the principle than that favoured by certain environ-mental groups (see above) and even that which appears in the 1990 North Sea Declaration. But then again, few politicians would feel confident enough to gamble resources today for uncertain benefits in the future or seek to curb economic 'progress' or individual freedoms on the basis of unproven risks of environmental damage. Nonetheless, one could argue that 'precipitate action' to safeguard the environment in advance of scientific proof of harm is one of the core premises of the precautionary principle. This is the kind of political and moral dilemma which underlies the simple notion that policies which relate to the environment should be more cautious. Third, in linking the application of precautionary measures to a 'comparison of likely costs and benefits' the UK is clearly laying down a marker that the benefits side of cost benefit analysis should be rea-sonably secure. For Britain, the line is that the benefit curve should be credibly underpinned by science or authoritative judgement. For the more environmentally aggressive ('leader') countries of the EU such as Germany, the Netherlands and Denmark, it is probably fair to say the climate of public opinion is more willing to accept that indeterminacy rather than science, should influence the shape of the curve.

In sum, if precautionary measures have to be justified by a comparison of costs and benefits and then only on the basis of 'sound science', then it is perhaps more accurate to describe the UK's overall philosophy as 'pre-ventative' rather than 'precautionary' (Cameron and Abouchar, 1991: 17).

THE APPLICATION OF THE PRECAUTIONARY PRINCIPLE IN THE UK

In this, the penultimate section, we consider a number of circumstances in which the UK government has attempted to implement, or has been forced to implement, precautionary measures. The case studies relate to four of the 'core elements' which we outlined above. They underscore the complex nature not only of the precautionary principle, but of environmental politics and policy making generally.

1   Preventative Controls at Source: The 'Red List' Saga

The UK has often approached international negotiations with a dogged determination not to take actions that appear economically costly, that run counter to the British philosophy of environmental regulation, or that appear to be based on spurious science. This negotiating stance was evident in the complex and often acrimonious debate between the European Commission and the UK Government over the discharge of dangerous substances to water, a full account of which has been given by Haigh (1989). The debate is particularly salient, because it was only resolved when the UK made important concessions consonant with the principle of precaution.

The debate turned on the most efficient, equitable and effective means to control the discharge of the most dangerous substances to water (the so-called 'black list'). The European Commission, pushed by the Dutch and the Germans, advocated uniform, source based emission standards tuned to what the 'Best Available (abatement) Technology' (BAT) could deliver. Meanwhile, the British Government, with the firm backing of British industry, was adamant that standards should be set in relation to what the local aquatic environment could bear. The first stance is technology based and is more compatible with the principle of *Vorsorge* than the second which is consistent with the 'dilute and disperse' paradigm. The conflict rumbled on for over a decade and was only really resolved when the UK, again under pressure from other European states, agreed to apply source-based BATNEEC (BAT not entailing excessive costs) controls to a certain number of substances on the Commission's list (the UK Government referred to these as its 'red list': ENDS, 154: 3). The Government's decision, announced in November 1989, represented a significant shift in not only the substance, but also the philosophy of UK environmental policy. For the first time, the UK accepted the logic of applying preventative, precaution-based, controls to the emission of the most potentially harmful substances to water, in both spirit and substance. BATNEEC-type controls had been applied to certain emissions to air since the passage of the first Alkali Act, but an important political precedent had still been set nonetheless. By making this concession, the UK Government agreed to reduce its net emissions of red list substances by 50 per cent between 1987 and 1995, irrespective of the harm they create or the capacity of the environment to absorb them. Even then, the red list does not cover all the substances on the Commission's black list and critics maintain that the government's approach is still 'entirely inconsistent with the precautionary approach' (Cameron and Abouchar, 1991: 9).

Finally it is not at all clear whether the government had actually implemented a true assessment of the costs of cutting emissions when it made its 1987 concession. Over a year later, the DoE had still not provided a firm estimate of the economic costs of complying with the terms of the agreement (DoE, 1988). In 1990, the government took a further step (but this time under its own initiative), when it announced the creation of a system of integrated pollution control which extends BATNEEC controls to certain solid waste arisings (Jordan, 1993).

2 Safeguarding Ecological Space: The Politics of Nature Conservation

The more strict, precautionary based, economic interpretations of the concept of sustainable development, emphasise the importance of protecting stocks of 'critical natural capital' (i.e., ecosystems, ecological processes, and natural resources that have no substitutes) because they are special, irreplaceable or valued highly. This is radical, since the orthodox economic view is that all forms of capital, be they natural or human made, are infinitely substitutable (Solow, 1986): losses of natural capital can be made up by technical substitution and innovation. Critical capital would include all 'keystone species or keystone processes' (Turner, 1993), such as the stratospheric ozone layer, tropical rainforests, important wetlands and ancient woodlands. In an intellectual sense, if one accepts the notion that there are stocks of critical capital which must be preserved intact (i.e., that cannot be substituted), then one has implicitly accepted that there are 'limits to growth' or 'environmental capacities' which should not be transgressed, and that in some instances it might be impossible or undesirable to attempt to reconcile development and environment interests as supported by the concept of sustainable development. If so-called 'sustainability constraints' (Jacobs, 1991) do exist and if they can be identified, then the precautionary principle suggests that they should not be breached (i.e., certain features of the environment should be treated as inviolable), *regardless* of any predicted costs and benefits (Jacobs, 1993). The designation of specific 'limits to growth' (or 'ecological buffers') is important because it precludes the application of cost/risk-benefit analysis and other forms of 'rational' assessment. In the case of critical assets, the precautionary line is to disallow development now and for ever more. This is certainly dissonant with the neo-classical economic paradigm and it presents 'a real challenge not only to the traditional presumption in favour of development, but to the utilitarian concept of balance which has been a fundamental tenet of the [UK] planning system' (Owens, 1994).

Whether the UK Government is actually doing enough to preserve stocks of critical capital is by no means clear, not least because there is little consensus about what is 'critical' and what is not. It has, however, made the following, somewhat intriguing, assertion: '[t]o some extent it is inevitable that economic development will affect the environment ... This is acceptable when the benefits outweigh the environmental costs but there are times when a feature of the environment needs to be treated as inviolable' (DoE, 1993: 10).

Unfortunately, the DoE failed to offer an example of what it meant by an 'inviolable' asset. In fact, there is very little agreement on which features of the environment should be designated as critical capital, or what criteria should be used to inform such a choice (Cowell, 1992). Owens (1994) rightly observes that the designation process will be politically contentious irrespective of which yardstick is used. In the UK, the stock of critical capital would certainly have to include SSSIs, Special Areas of Conservation (SACs), Areas of Outstanding Natural Beauty (AONBs) and so on. And what is the UK's record on protecting these important ecological assets? Of the 5852 SSSIs in 1992, seven had been lost and 313 were experiencing long or short term damage caused by agricultural practices, acidification and tourism (Brown, 1993). Ratcliffe (1993) documents the threats facing some of the UK's most important ecological sites, and questions whether the Government is conserving enough to implement the EC's Habitats Directive. In 1993, English Nature (1993: 5) reported that: 'Society's past failure to adopt the principle of sustainable land use has resulted in the national decline of *most* of England's wildlife and natural features. It has led to the local extinction of many species, not just the rare, but also the commonplace' (emphasis added).

The first comprehensive survey of the state of the British countryside during the 1980s was published by the Government in 1993. It catalogued the progressive and widespread loss of natural habitats, biodiversity, deciduous trees and riverbanks to activities such as quarrying, road building, urban sprawl and agriculture (Pearce, 1993). This is certainly not consistent with some of the stronger formulations of the precautionary principle, which emphasise the non-instrumental value of non-human entities.

3   Acting in Advance of Scientific Proof: The Ban on Sewage Sludge Disposal at Sea

At the 1987 North Sea Conference, the UK, under intense pressure from other European states, accepted that restrictions would need to be placed on the dumping of sewage sludge at sea, and at the third (Hague) meeting three years later (1990), it agreed to terminate all disposal to sea by 1998. The move was justified on the grounds of precaution. The same might also be said to apply to the government's 1994 confirmation that it would comply with the international moratorium on the dumping of radioactive waste at sea.

Prior to 1987, the UK had supported the dumping of sewage sludge at sea on the basis that it was the best practicable environmental option (BPEO: see MAFF, 1990), although some doubted whether the Government had actually conducted a thorough assessment of the relative costs and benefits associated with the various disposal options (land based incineration, composting, etc.: Mayer, 1987). With the foreclosure of the sea disposal route, the water companies are now searching for alternatives, all of which generate a complicated melange of social costs and benefits. Clark (1991), for example, warns that the costs associated with the alternative methods of disposal may actually be greater than dumping waste at sea.

In a UK context, it was later estimated that the ban on dumping would require five of the privatised water companies to spend £100m on additional capital expenditure and £0.4m per annum thereafter (Water Research Centre, or WRC, 1992). The WRC has suggested that, with the foreclosure of the sea route, the BPEO is to incinerate the sludge on land. There are, inevitably, disamenity effects associated with this route (air pollution and the disposal of ash) as with every other. Ironically, the implementation of EU Directive 91\272 (Urban Wastewater Treatment) will actually result in a significant increase in sludge production in the future.

From a science or economics perspective, the decision to ban the dumping of sludge at sea was a politically expedient and poorly informed action. For substances of low toxicity like sewage sludge, it might be better to proceed by way of two steps: the first cautionary, such as more research, some extra controls and carefully targeted analyses (e.g., multi-criteria decision analyses: Voogd, 1983), or some form of integrated hazard assessment procedure (Gray *et al.*, 1991: 436); the second precautionary, but only when it is warranted. Careful monitoring over time would alert policy makers to any undesirable environmental effects.

4   Paying for Ecological Debt: The Strict Liability Debate

Strict liability is already provided for in common law and aggrieved parties
can also resolve disagreements and negotiate compensation arrangements
informally out of court. However, it is the steady incorporation of strict liability clauses into statute law that is causing industry and the insurance
markets the most concern. Ball and Bell (1992: 131) give examples of UK
statutes that impose liability by means of private law remedies. For
example, section 34 of the EPA provides for a statutory duty to be imposed
on any person who produces, carries or disposes of waste or, as a broker,
has control of waste. Any person to whom the duty applies should take reasonable steps to, *inter alia*, prevent the escape of waste, and make sure that
the waste is only transferred to an authorised person. In effect, the responsibility to act in a precautionary mode is passed down the waste chain, from
source to eventual sink. The Water Resources Act 1991 enables the NRA to
clean up pollution and recover the relevant costs from the waste discharger,
although it has yet to make use of this facility (ENDS, 225: 19).

The European Commission has produced a Green Paper on liability and
compensation for environmental damage and is preparing to legislate to
create a strict liability regime coupled to compensation funds to pay for
restoration where a liable party cannot be found or is unable to pay. The
Commission is concerned that competition between member states may
be hindered if each has its own particular liability regime. Both the UK
Government and industry have voiced strong opposition to the introduction of liability measures at the EC level. Industry is particularly concerned about the high costs of remediation or compensation, and the (CBI)
has made it known that British industry is strongly opposed to the retrospective application of strict liability (ENDS, 225: 19). In March 1993,
the Government abandoned plans to implement the section of the EPA
which merely provides for the registration of contaminated land, in the
face of strong pressure from property developers concerned about 'planning blight'. In this last context, the CBI has declared that remedying con-
tamination, the cost of which they estimate to be up to £40 billion, may be
the biggest challenge facing industry in the 1990s (*Guardian*, 3 November
1993). This is a good example of how domestic policy priorities based on
maintaining economic dominance of key aspects of industrial growth can
overwhelm a precautionary mode. In international terms Britain may
accede to precautionary principles, as in sludge and low-level radioactive
waste dumping. But purely at the domestic level, the government can, and
does, readily eschew the principle if it threatens other political priorities.
The government, for example, has made it clear that it will not ratify an

existing Council of Europe Convention on Civil Liability caused by dangerous activities.

Meanwhile, in common law, a potentially important UK precedent on the retrospective application of liability was set by the Appeal Court's judgement in *Eastern Counties Leather* v. *Cambridge Water Company*. The case centred on the discharge of an organochlorine solvent from leaks in containers stored at the premises of the tanning company. In the event, the solvent entered the groundwater to seep into a borehole supplying potable drinking water over a period of almost 30 years. The concentrations subsequently breached the EC Drinking Water Directive, so the source of supply was discontinued subject to dilution. The Appeal Court held that the tanning company was strictly liable under the tort of nuisance, though not under the normal strict liability doctrine (ENDS, 237: 41). However in an historic judgement, the House of Lords overturned this judgement and held that no reasonable operator could have foreseen that small spillages could have resulted in pollution 5 km away. Their Lordships dismissed the liability charge, but they did note that regulatory guidelines and enforcement measures should be more prevention conscious so that the civil law should not have to be in a position of defining anticipatory liability (ENDS, 227: 43–4).

This judgement places the spotlight on the powers currently available to the custodial agencies to be precautionary in the face of both data deficiency and ignorance when contemplating the protection of critical natural capital or removing a potentially undesirable genetically modified organism, or establishing compensation principles for the restoration of land contaminated by industrial activity long gone.

CONCLUSION

The precautionary principle is vague enough to be acknowledged by all governments regardless of how well they protect the environment. But the politics of precaution are also powerful and progressive and, in the context of this book, they offer a profound critique of many of the ways in which the UK Government has managed its environmental affairs. Wrapped up in the debate about precaution are forceful new ideas which point the way to a more preventative, source based, integrated and biocentric basis for policy. The politics of precaution are also the politics of a more international environmental policy process and resonate within a society that is more risk averse. Ultimately, domestic public pressure and international obligation will be the principal forces driving the wider application of precaution in the UK. Only when precaution tips the balance in favour of environmental well-being rather than material growth will the UK begin to make purposive steps in the direction of sustainable development. Currently, the British Government seems prepared to toe the diplomatic line when pushed by other states (as in the case of dumping sewage sludge and, more recently, radioactive waste at sea), but it has, heretofore, shown little inclination to lead the international community on environmental matters. The emphasis which the government places on 'sound science' and careful cost-benefit analysis, suggests that, at heart, it is still highly suspicious of the precautionary principle and wary of the threat which it appears to hold to economic growth and 'rational' policy making.

Our contention is that precaution will not explode on to the environmental stage, sweeping away all forms of risk or cost benefit assessment, careful scientific analysis and existing legal norms relating to the relative power of polluters and victims. Rather, it will seep through the pores of decision making institutions and the political consciousness of humanity by stealth. It will do this when, and if, it has the tide of the times behind it.

Notes

1.   This is consistent with the practice of setting a 'safe minimum standard' to protect important features of the environment (Ciriacy-Wantrup, 1963; Bishop, 1978).
2.   Points beyond which irreversible phase changes may occur, or ecosystems may collapse.
3.   See, for example, the 1990 Ministerial Declaration on the North Sea.
4.   See, for example, the 1990 Bergen Ministerial Declaration on Sustainable Development.
5.   Germany's environmental expenditure, judged as a percentage of GDP, is one of the highest amongst OECD countries, and the boost this has given to eco-industrial sectors has been considerable. In 1992, the OECD estimated that the German eco-industry as a whole employed 320 000 people and had a gross turnover of DM 40m (OECD, 1992).
6.   The others were: using the best scientific evidence available, public information, international cooperation, and use of the best policy instruments.

References

Ball, S. and Bell, S. (1992), *Environmental Law* (London: Blackstone Press).
Barker, A., Peters, B. and Guy, C. (1993), *The Politics of Expert Advice: Creating Using and Manipulating Scientific Knowledge* (Edinburgh: Edinburgh University Press).

Bennett, G. (1992), *Dilemmas: Coping With Environmental Problems* (London: Earthscan).

Bishop, R. C. (1978), 'Economics of Endangered Species', *American Journal of Agricultural Economics*, 60, 10–18.

Bodansky, D. (1991), 'Scientific Uncertainty and the Precautionary Principle', *Environment*, 33, 4–5 and 43–5.

Boehmer-Christiansen, S. (1988), 'Black Mist and the Acid Rain: Science as a Fig-Leaf Of Policy', *Political Quarterly*, 59, 145–60.

Boehmer-Christiansen, S. and Skea, J. (1991), *Acid Politics: Environmental and Energy Policies in Britain and Germany* (London: Belhaven Press).

Boehmer-Christiansen, S. (1992), 'Anglo-German Contrasts in Environmental Policy Making', *International Environmental Affairs*, 4, 295–322.

Boehmer-Christiansen, S. (1994), 'The History and Application of the Precautionary Principle in German Environmental Policy', in T. O'Riordan and J. Cameron (eds), *The Precautionary Principle in Environmental Policy* (London: Cameron and May).

Boulding, K. (1966), 'The Economics of the Coming Spaceship Earth', in H. Jarrett, *Environmental Quality in a Growing Economy* (Baltimore, MD: Johns Hopkins University).

Brown, K. (1993) 'Biodiversity', in D. W. Pearce, R. K. Turner, T. O'Riordan, N. Adger, G. Atkinson, I. Brisson, K. Brown, R. Dudourg, S. Frankhauser, A. Jordan, D. Maddison, D. Moran, J. Powell, *Blueprint 3* (London: Earthscan).

Brown, K. and Jordan, A. (1993), 'The UK Government's Response to the UNCED', *Pacific and Asian Journal of Energy*, 3, 87–99.

Cameron, J. and Abouchar, J. (1991), 'The Precautionary Principle', *Boston College International and Comparative Law Review*, 14, 1–27.

Cameron, J. and Wade-Gery, W. (1992), *Addressing Uncertainty: Law, Policy and the Development of the Precautionary Principle*, CSERGE Working Paper GEC 92-43 (UCL and UEA: CSERGE).

Ciriacy-Wantrup, S. V. (1963), *Resource Conservation* (Berkeley, CA: University of California Press).

Clark, R. B. (1991), 'Assessing Marine Pollution and its Remedies', *South African Journal of Marine Science*, 10, 341–51.

Cowell, R. (1992), *Take and Give*, United Kingdom Centre for Economic and Environmental Development (UKCEED) Discussion Document 10 (Cambridge: UKCEED).

Dobson, A. (1990), *Green Political Thought* (London: Harper Collins).

DoE (1988), *Inputs of Dangerous Substances to Water: Proposals for a Unified System of Control* (London: Doe).

DoE (1993), *UK Strategy for Sustainable Development* (London: DoE).

Dunlap, R. E., Gallup, G. H. and Gallup, A. M. (1993), 'Of Global Concern: Results of the Health of the Planet', Survey, *Environment,* 34, 6–14.

Eckersley, R. (1992), *Environmentalism and Political Theory* (London: UCL Press).

English Nature (1993), *Strategy for the 1990s*, Consultation Paper (Peterborough: English Nature).

Environmental Data Services (1987) 'UK Policy shifts on Dumping' *ENDS Report*, 154 (November) 3.

Environmental Data Services (1993) 'Opening Shots in the Liability Debate' *ENDS Report* 225 (October) 19–20.

Environmental Data Services (1993) 'Key Ruling on Civil Liability by House of Lords' *ENDS Report* 227 (December) 43–4.

Environmental Data Services (1994) 'Precautionary Principle not binding in English Law' *ENDS Report*, 237 (October) 41–42.

Freestone, D. (1991), 'The Precautionary Principle' in R. Churchill and D. Freestone (eds), *International Law and Global Climate Change* (London: Graham & Trotman).

GESAMP (1986), *Environmental Capacity: An Approach to Marine Pollution Prevention*, Food and Agriculture Organisation Report No. 30 (New York: Food and Agriculture Organisation).

Gleick, J. (1987) *Chaos* (New York: Viking Penguin).

Gray, J. S. (1990), 'Statistics and the Precautionary Principle', *Marine Pollution Bulletin*, 21, 174–6.

Gray, J. S, Calamari, D., Duce, R., Portmann, J. E., Wells, P. G., and Windom H. L. (1991), 'Scientifically Based Strategies for Marine Environmental Protection and Management', *Marine Pollution Bulletin*, 22, 432–40.

Haas, P. (1990), *Saving the Mediterranean: The Politics of International Environmental Cooperation* (New York: Columbia University Press).

Haigh, N. (1989), *EC Environmental Policy and Britain,* 2nd edn (London: Longman).

Haigh, N. (1994), 'The Precautionary Principle in British Environmental Policy', in T. O'Riordan and J. Cameron (eds), *The Precautionary Principle in Environmental Policy* (London: Cameron & May).

Hey, E. (1992), 'The Precautionary Principle in Environmental Law and Policy: Institutionalising Precaution', *Georgetown International Environmental Law Review*, 4, 303–18.

HM Government (1990), *This Common Inheritance: Britain's Environmental Strategy.* (London: HMSO) Cmnd 1200.

HM Government (1994), *Sustainable Development: The UK Strategy* (London: HMSO) Cmnd 2426.

Horsman, P. (1992), 'Reduce it; Don't Produce it: The Real Way Forward' in T. O'Riordan and V. Bowers (eds), *IPC: A Practical Guide for Managers* (London: IBC Technical Services).

Inglehart, R. (1989), *Culture Shift* (Princeton, NJ: Princeton University Press).

Jacobs, M. (1991), *The Green Economy* (London: Pluto Press).

Jacobs, M. (1993), *Sense and Sustainability* (London: CPRE).

Jasanoff, S. (1990), *The Fifth Branch: Science Advisers as Policy Makers* (Cambridge, MA: Harvard University Press).

Johnston, P. and Simmonds, M. (1990), 'Precautionary Principle', *Marine Pollution Bulletin*, 21, 402.

Jordan, A. J. (1993), 'Integrated Pollution Control and the Evolving Style and Structure of Environmental Regulation in the UK', *Environmental Politics*, 2, 405–27.

Jordan, A. J. and O'Riordan, T. (1993), 'Implementing Sustainable Development: The Political and Institutional Challenge', in D. W. Pearce *et al., Blueprint 3* (London: Earthscan).

Keohane R., Haas, P. and Levy, M. (1993), *Institutions for the Earth* (Cambridge, MA: MIT Press).

Lean, G. (1991), 'The Role of the Media', in L. Roberts and A. Weale (eds), *Innovation and Environmental Risk* (London: Belhaven).

Liefferink, J. D., Lowe, P. and Mol, A. (eds) (1993), *European Integration and Environmental Policy* (London: Belhaven Press).

Lorenz, E. (1993), *The Essence of Chaos* (London: UCL Press).

Lowe, P. and Flynn, A. (1989), 'Environmental Politics and Policy in the 1980s', in J. Mohan (ed.), *The Political Geography of Contemporary Britain* (London: Macmillan).

MAFF (1990), *The Utility of Experimental Measures of Biological Effects for Monitoring Marine Sewage Sludge Disposal Sites*, Aquatic Environment Monitoring Report Number 24 (London: MAFF).

Maxwell, J. and Weiner, S. (1993), 'Green Consciousness or Dollar Diplomacy?: The British Response to the Threat of Ozone Depletion', *International Environmental Affairs*, 5, 19–41.

Mayer, M. (1987), *Implementing BPEO: Some Opportunities and Constraints*, paper presented at the Institute of Wastes Management Conference, London, 8 May 1987.

McLoughlin, J. and Forster, M. (1981), *The Law and Practice Relating to Pollution Control in the United Kingdom*. (London: Graham & Trotman).

McGarvin, M. (1994), 'The Precautionary Principle and Ecological Science', in T. O'Riordan and J. Cameron (eds), *Interpreting the Precautionary Principle* (London: Cameron & May).

McGrew, A. (1993), 'The Political Dynamics of the "New" Environmentalism', in D. Smith (ed.), *Business and the Environment* (London: Paul Chapman).

OECD (1992), *OECD Environmental Performance Reviews: Germany* (Paris: OECD).

O'Riordan, T. (1992), *The Precaution Principle in Environmental Management*, CSERGE Working Paper GEC 92-03

O'Riordan, T. (1993), *Interpreting the Precautionary Principle*, CSERGE Working Paper PA 93-03 (UCL and UEA: CSERGE).

O'Riordan, T. and Cameron, J. (1994), 'The History and Contemporary Significance of the Precautionary Principle', in T. O'Riordan and J. Cameron (eds). *The Precautionary Principle in Environmental Policy* (London: Cameron & May).

Owens, S. (1994), 'Planning and Nature Conservation: the Role of Sustainability', *ECOS*, 14, 3–4, 15–22.

Pearce, D. W., Turner, R. K. and O'Riordan, T. (1993), 'Energy and Social Health: Integrating Quantity and Quality in Energy Planning', *Energy and Environment*, 4, 155–173.

Pearce, D. W. (1994), 'The Precautionary Principle and Economic Analysis,' in T. O'Riordan and J. Cameron (eds), *The Precautionary Principle in Environmental Policy* (London: Cameron & May).

Pearce, F. (1993), 'Urban Sprawl Devours Britain's Countryside', *New Scientist*, 27 November, 8.

Porras, I. (1993), 'The Rio Declaration', in P. Sands (ed.), *Greening International Law* (London: Earthscan).

Ratcliffe, D. (1993), *Conservation in Britain: Will Britain Make the Grade?* (London: Friends of the Earth).

Rehbinder, E. (1988), 'Vorsorgeprinzip Umweltrecht und Preventive Umweltpolitik', in U. Simonis (ed.), *Preventive Umweltpolitik* (Tokyo: United Nations University Press).

Rowlands, I. (1993), *The Fourth Meeting of the Parties to the Montreal Protocol: Response and Reflection,* CSERGE Working Paper 93-18 (UCL and UEA: CSERGE).

Royal Commission on Environmental Pollution (1984), *Tenth Report: Tackling Pollution: Experience and Prospects* (London: HMSO) Cmnd. 9149.

Solow, R. M. (1986), 'On the Intergenerational Allocation of Natural Resources', *Scandinavian Journal of Economics*, 88, 141–9.

Stebbing, A. R. D. (1992), 'Environmental Capacity and the Precautionary Principle', *Marine Pollution Bulletin*, 24, 277–95.

Stone, C. (1987), *The Earth and other Ethics: The Case for Moral Pluralism* (New York: Harper & Row).

Tait, J. and Levidow, L. (1992), 'Proactive and Reactive Approaches to Risk Regulation: The Case of Biotechnology', *Futures*, April, 219–231.

The Economist (1993), 'The Environment: Turning Brown', 3 July.

Turner, R. K. (1993), 'Sustainability: Principles and Practice', in R. K. Turner (ed.), *Sustainable Environmental Economics and Management* (London: Belhaven Press).

Von Moltke, K. (1988), 'The Vorsorgeprinzip in West German Environmental Policy', in Royal Commission on Environmental Pollution, *Twelfth Report: Best Practicable Environmental Option* (HMSO: London) Cmnd 310.

Von Moltke, K. (1991), Three Reports on German Environmental Policy, *Environment*, 33, 7, 25–9.

Voogd, H. (1983), *Multi-Criteria Evaluation for Regional and Urban Planning* (London: Pion).

Weale, A. (1992), *The New Politics of Pollution* (London: Manchester University Press).

Weale, A. (1993), 'Ecological Modernisation and the Integration of European Environmental Policy', in J. D. Liefferink, P. Lowe and A. Mol (eds), *European Integration and Environmental Policy* (London: Belhaven Press).

Weinberg, A. M. (1985), 'Science and its Limits', *Issues in Science and Technology*, 2, 59–72.

Wilkinson, D. (1992), 'Maastricht and the Environment', *Journal of Environmental Law*, 4, 221–39.

WRC (1992), *A Methodology for Undertaking BPEO Studies of Sewage Sludge Treatment and Disposal* (Medmenham: WRC), December 1990.

Wynne, B. (1992), 'Uncertainty and Environmental Learning: Reconceiving Science in the Preventative Paradigm', *Global Environmental Change*, 2, 111–27.

Wynne, B. and Meyer, S. (1993), 'How Science Fails the Environment', *New Scientist*, 5 June, 33–5.

6 UK Environmental Policy and the Politics of the Environment in Northern Ireland in the 1990s

Steven Yearley

INTRODUCTION

There are two reasons for including a chapter on Northern Ireland in this volume on UK environmental policy in the 1990s. The first has to do with the inherent interest of examining environmental action and legislation in a politically divided and economically struggling area. The familiar arguments about environment versus jobs take on a new salience in this context, while the challenges of sustainable development seem all the greater in a situation where even political stability may not be sustainable. Second, Northern Ireland can stand as an example – albeit a stark one – of regional differences within the UK. Such differences may well influence the appropriateness of, and the local responses to, environmental policies, which are generally devised on a UK-wide basis. Although the details of a Welsh or Scottish perspective on UK environmental policy would differ from the present account, this chapter will serve to indicate that centralised UK environmental priorities, targets and practices may not appear to be as reasonable or compelling at the periphery as they do at the centre. The reasons for centralisation, both political and ideological on the part of the UK Government and administrative on the part of the EU, are explained in other chapters. The Northern Irish case offers a perspective from the periphery.

THE NORTHERN IRISH CONTEXT

The context for environmental policy in Northern Ireland is shaped by five main factors. Without implying that this is the most important factor, we can reasonably start with its peripheral geographical location and the condition of the natural environment itself. Northern Ireland has around 6 per cent of the land area of the UK with less than 3 per cent of the population. Agriculture is accepted as the single most important industry in Northern Ireland, involving around 8 per cent of the population; agriculture makes three times as large a contribution to the Northern Irish economy as it does in Great Britain. It is thus a conspicuously rural region characterised by small-scale agricultural production. There are few large farms, certainly: in the words of Seamus Heaney, 'we have no prairies', not even of the sort found in the east of England. On average, farm sizes are rather less than half those in England. In line with this economic orientation, outside a few large urban areas air quality is good, despite the widespread use of coal for domestic heating. Unless polluted by farm wastes, the majority of rivers tend to be clean and there is relatively little industry to cause pollution. Officials' boasts about NI's green image have some basis in fact. However, NI's small scale and physical and political isolation have not been uniformly beneficial to the environment. Connection to infrastructural facilities, including those such as North Sea gas which would have benefited the environment, has not been encouraged, and the region's small size has led to diseconomies of scale in, for example, electricity generation. Energy efficiency is poor.

It is the very lack of industry along with the 'Troubles', for which NI is internationally famous, that provides the second contextual element. In practice, security policy and the pursuit of economic regeneration have tended to trump all other policy considerations. This was made explicit in the 1990 report of the House of Commons Environment Committee (henceforth the Rossi Report) where Peter Bottomley, then minister responsible for most aspects of environmental policy, acknowledged, that 'In many places in Northern Ireland if one person wants to put a brick on top of another I would be tempted to pay for them to do it' (1990: 103). The potential for a trade-off between jobs and environmental protection has been a consistent theme in the Northern Irish context.

It is important to realise, moreover, that the North's lack of (polluting) industry is relatively recent; even into the 1960s there was mass employment in the engineering, textile and shipbuilding sectors (see Jefferson, 1987; Harris, 1990). Unlike the Irish Republic where there are large areas which have never experienced industrialisation at all, NI has experienced de-industrialisation. There is a strong desire to pursue policies aimed at fostering economic development both for the sake of countering unemployment and because this is considered to be an aspect of security policy, since the perpetuation of the political unrest and violence is associated by many with the lack of job prospects. Furthermore, the industrial past has bestowed an environmental legacy since some of the departed industries have left behind contaminated sites and, given the looser legislation in force at the time of their departure, little may be known about such residual waste (see Rossi Report, 1990: 35 and 40).

In the light of this industrial decline the UK Government, even in its most free-market periods, has effectively pursued interventionist policies in the North of Ireland. This has taken a variety of forms, including numerous job-creation schemes, often futile attempts to attract overseas investment and the encouraging of the relocation of certain services (for example, the Directory Inquiries service) to Northern Ireland. Additionally, the role of the government as a direct employer has grown sharply. This is because there are high numbers of jobs in social services (in response to unemployment and deprivation) and in the security forces. Some commentators have even suggested that the growth of security industry jobs (not only the police, but shopping mall guards and suppliers of surveillance and so on) has actually produced more jobs than have been lost owing to the disruption of the 'Troubles'; in this limited sense the Troubles may even have been good for employment prospects.

In any event, the result has been that while the private sector has declined, the public sector has grown; thus, the proportion of workers employed by the state and the percentage of economic activity attributable to it are very high by Western European standards. Consequently, there is much criticism of the way that the Northern Irish environment gets treated. The state regulates pollution and enforces those regulations; at the same time it promotes industry and is the leading employer and principal economic force.

Despite this highly influential position, governmental agencies and ministers are curiously unaccountable and, in a significant sense, politically dependent. New legislation comes to Northern Ireland by a circuitous route which delays its implementation; the time lag is reckoned to be at least 64 weeks (Rossi Report, 1990: xiv). Legislation for England and Wales cannot be applied directly in NI because the existing framework of laws and institutions is somewhat different and because, since Direct Rule was introduced in 1973, the distribution of power between central government and local authorities has deviated significantly from the situation in Great Britain. Regarded as unprofessional and often sectarian, local authorities had been under review since 1969. In 1972 they were finally stripped of most of their responsibilities for housing allocation, education and health (Elliott and Wilford, 1987: 293–4), so that twenty-six small councils now preside over residual tasks summarised as 'bins, bogs and burials'. The maintenance of a separate system of NI law is also held to be 'a good thing' as it would facilitate devolution if new political structures were agreed.

At the same time, locally elected MPs do not belong to any of the dominant UK parties and are thus not appointed as NI ministers; furthermore, they are never more than sporadically influential in the House of Commons. Only in late 1993 was there even agreed a Select Committee for Northern Ireland. Accordingly, the responsible ministers are only tenuously related to the voters of Northern Ireland.

The fifth influence, and one of growing importance, is exerted by the EU. NI is an Objective One Region and is thus eligible for a great variety of development funding. As has been documented elsewhere (see the essays in Baker, Milton and Yearley, 1994) the EU tends to have contradictory influences on peripheral regions. To some degree, European legislation and the existence of EU institutions, to which appeal can be made, act as a 'progressive' force. In other words, they tend to raise legal and (in most cases) practical standards on matters such as environmental protection towards the European average. On the other hand, the amount of resources going to these objectives is dwarfed by the sum devoted to promoting economic development in the periphery, which often means road construction and other projects inimical to environmental protection.

Given this background, the remainder of this chapter is concerned with reviewing how the environmental policy performance of the UK has been reflected in and affected NI.

THE APPARATUS FOR ENVIRONMENTAL POLICY IN NORTHERN IRELAND

As explained above, the leading characteristic of environmental policy (in the narrow sense) in NI is the way it lags behind policy in Great Britain on account of parliamentary delays. At the time of the Rossi Report, the DoE NI listed 33 outstanding legislative commitments (1990: 206–10). Reasonably enough, laws are seldom different; one can think of no acceptable grounds for accepting, say, a higher local tolerance for air pollution in NI than in Runcorn or Lincoln. There are one or two instances where laws would apply, for example, to so few plants that the legislative drafting seems never to have taken place (see 1990: 106 for an illustration). But the main issue is that legislative changes (which have generally been in the direction of raising environmental standards) have simply come later to NI than the rest of the UK. Occasionally laws have been shaped in a narrower way. For instance, there is to be no 'integrated pollution control' legislation *per se* in NI; instead the same standards as apply in Great Britain are being sought through the tightening of air pollution controls. Officially, this difference is justified by arguing that in the context of Northern Irish industry the same environmental quality can be achieved with simpler legislation, thus speeding the introduction of the new standards. But this is the exception. In the great majority of areas legislation mirrors the Great Britain precedent, a year or two later. Occasionally this can bring advantages. Ineffective legislation or poor drafting can be corrected before NI laws are enacted. But mostly the delay means simply that environmental problems persist longer in NI than elsewhere. These delays were a prime focus for criticism by the Environment Committee; in certain cases the delays have been so pronounced that they have led to the UK being in contravention of European timetables (1990: 105 and 211–12). Currently, for example, air quality in Belfast still runs the risk of breaching EU standards, despite the fact that all member states were to have complied with the Directive on smoke and sulphur dioxide by April 1993. This example, however, demonstrates the connection between environmental policy and other social and economic policies (such as the level of welfare benefits and the decision whether to supply piped-in gas) outside the control of DoE NI; it might therefore be regarded as an anomaly. I shall return to this issue later on.

The recent history of the enactment of Northern Irish environmental legislation is thus relatively straightforward, but the practical assessment of policy must also take monitoring and implementation into account. Both these topics have been subject to heavy criticism from local environmentalists. In particular, attention has focused on the slow progress of the designation of Areas of Special Scientific Interest (or ASSIs, Northern Ireland's version of SSSIs) and on the restricted amount of air quality monitoring: the principal monitoring station in Belfast is in a pedestrianised area. The officials responsible for environmental policy attribute these problems to understaffing and the lack of financial resources for the area. As elsewhere in the UK (see Richardson, Ogus and Burrows, 1982), there is also the issue that pollution control officials realise that they have to achieve *a modus vivendi* with the people whom they regulate. Indeed, in the government's reply to the Rossi Report (DoE NI, 1991: 12) it was explicitly stated that 'compliance is often achieved more effectively by consultation than by legal penalties'. The apparent preference for 'consultation' has led to disputes between officials and the voluntary sector over standards in the regulation of agricultural practices, independent waste disposal contractors and industrial polluters, with pressure groups claiming that officials are insufficiently robust in their regulatory activities. Insofar as several environmentalist groups are headed by 'blow-ins', while such polluters and their regulators are principally Northern Irish, there may be some special characteristics to this issue in NI. However, the main influence in policy is not this cultural conflict but institutional conflicts, which are described in the next two sections.

INSTITUTIONAL CONFLICTS AND THE ENVIRONMENT SERVICE

A long-running complaint of the voluntary sector has been that environmental protection has been delegated to government departments, not to independent or even 'quasi-independent' agencies.[1] During the 1980s this complaint was heard primarily in relation to wildlife conservation. Where Great Britain had the NCC, NI had no such 'independent' agency. The corresponding functions were carried out by a branch of the DoE NI. This arrangement was felt to have three drawbacks. First, the Countryside and Wildlife Branch was relatively unimportant within the DoE NI overall and thus received little money and staffing resources. Second, being small and lacking top officials of the highest grade, it could not have the same influence within the DoE NI as the Directors of the Roads or the Planning Services (Milton, 1990: 71). As evidence for these two propositions, environmentalists in the voluntary sector tend to point to examples including the Antrim Area Plan where it was, in effect, left to the Royal Society for the Protection of Birds (RSPB) to argue for the importance of wildlife considerations in the shaping of the land-use priorities for the area plan, an argument which the branch had apparently failed to make (Yearley and Milton, 1990: 199). Finally, being both within the civil service as a whole and within the DoE NI in particular, officials of the branch were not free to speak out about policy issues when their views conflicted with those of other divisions of the civil service. On occasions the NCC joined with voluntary sector bodies in Great Britain to oppose government policy or governmental decisions. In practice the branch could do no such thing. Even when pressure groups were informally convinced that branch officials were sympathetic to environmentalist objections, officials toed the civil service line (Milton, 1990: 74).

Three things have happened in the 1990s to modify and broaden this argument. First, the NCC was dismantled into separate bodies for England, Scotland and Wales. This had contrasting implications for the NI situation. On the one hand, the question naturally arose, if three UK territories can have extra-governmental wildlife and countryside protection bodies, why could not the fourth? However, the dismantling of the NCC (in the view of many environmentalists, as a punishment for being too 'independent' and critical of government) appeared to signal a loss of power for these bodies; having an NCC-like agency for NI no longer looked like such an attractive or significant prospect.

Second, in response to the 'greening' of the UK Government generally and to pressure from the Environment Committee and from local environmentalists (notably the Northern Ireland Environment Group (NIEG), an umbrella organisation representing wildlife and amenity conservation interests with good contacts with government), conservation concerns within the DoE NI were rationalised. A new Environment Service was formed with a Director sitting on the DoE NI directorship and with an internal subdivision into wildlife and heritage protection on the one hand, and environmental protection on the other. In 1989 a standing advisory committee (the CNCC) was established with responsibility chiefly for advising on wildlife and amenity issues; however, it is entitled to be legally represented at public inquiries and is requested to advise the DoE NI on grant aid for voluntary bodies so its responsibilities can be spread a little more widely than in its apparently narrow initial remit. In its first report, for 1989–90, the CNCC set out its intention to 'develop policies on environmental issues such as fish farming, lignite mining, solid waste disposal, etc. inasmuch as these activities have impact on landscape and wildlife' (1991: 35).

This seems to grant the CNCC considerable interpretative flexibility in determining what subjects it can properly speak on and which not. This impression was reinforced by the colourful testimony given to the Environment Committee by John Phillips (the Director of Conservation, then on the brink of retirement) who claimed that 'Both Mr Bottomley and his predecessor have said to [the CNCC], 'Look, if you want to go public and knock hell out of the Government, go ahead and do it because that is what you are there for' (1990: 87).

While stopping well short of the liberties offered by Phillips, the CNCC report was also critical of what it claimed was underfunding of the necessary work of the Countryside and Wildlife Branch in, for example, ASSI designation. Such statements by the CNCC and the reforms within the DoE NI's structure indicated to commentators and the Northern Irish voluntary sector that environmental interests within government had been strengthened. Maybe, it was felt, this internal reform would rectify the majority of problems of which environmentalists had complained. However, and this was an issue pursued at length by the Environment Committee, the CNCC has virtually no staff of its own. It is promised technical support from DoE NI but faces a potential conflict of interest over such support since, presumably, the more demands it places on technical staff, the less time those staff will have to do the survey and regulatory work of the Department. Despite a clear recommendation from the Environment Committee the government declined the proposal for a separate professional staff for the CNCC (DoE NI, 1991: 23). Supporters of environmental NGOs are also aware that many members of the CNCC receive detailed information about environmental issues (officially at least) exclusively from the DoE NI; discrete lobbying aside, they may never know of the alternative policy proposals of the voluntary agencies or even, for that matter, of the farming or business sectors. Accordingly, there is room for doubt about the extent to which the CNCC can be an effective and independent check on the environmental work of the DoE NI.

Third, partly as a result of water and electricity privatisation in Great Britain and partly because of environmental reforms pursued by central government, a number of independent monitoring bodies came into existence and/or prominence in Great Britain, including the NRA, the Drinking Water Inspectorate and HMIP. There has also been a (waxing, waning and re-waxing) expectation that these watchdogs would coalesce into an independent Environmental Protection Agency. Voluntary-sector environmentalists in NI have been uncertain how to respond to this development and to the model for NI it implicitly sets out. While the Rossi Report was critical of the arrangements for environmental protection in NI and favoured the setting up of an independent NI Environment Protection Agency (though more as an executive agency than as a watchdog), the Committee's own published 'Recommendations' implied that this might not be deemed feasible and thus offered the government an immediate 'escape route' from the proposal (1990: xxxv). During the inquiries of the Environment Committee the minister and senior officials had argued for the benefits of an in-house environmental protection service, claiming both that it would be effective in the small Northern Irish community and well placed to hear of injurious proposals before they gained too much support. Environmental campaigners are accordingly unsure whether plans for an NI Environmental Protection Agency would ever be realisable without exceptional new sources of political support. In any case, there is a concern about the extent to which any such type of body could be independent and effective in the Northern Irish context. With a small civil service and tiny governmental elite, any such agency might be further from the source of power than the present Environment Service and might thus be ineffective. In particular, given the official emphasis on economic and security policy and the fact that the most recent minister (Tim Smith) and both his predecessors (Robert Atkins and Richard Needham) had responsibility for the economic development *and* environment briefs, there have been doubts whether any NI Environmental

Protection Agency's 'independence' would amount to much. Those who do still entertain hopes of some form of environmental agency for NI now tend to favour a watchdog-only model, roughly comparable to the Fair Employment Agency, which aims to combat sectarian discrimination in employment. The creation of any such body would in all probability be opposed by the Environment Service (see the tone of comments from the DoE NI, in House of Commons Environment Committee 1993, 17–8).

ECONOMIC AND ENVIRONMENTAL OBLIGATIONS IN THE NI CONTEXT

The first institutional influence on environmental policy in NI is plainly the make-up and position of the Environment Service itself. It is questionable how independent and zealous the Environment Service has been able to be. It is also unclear whether a fully institutionally independent Environmental Protection Agency is a realisable next step. But the second influence on environmental policy comes from the broader context within which environmental policy has to be made. The economy of NI is in a woeful condition: it costs the UK exchequer around £2 billion each year (Bradley, 1990: 447), and the high rates of unemployment contribute to the feelings of hopelessness which run parallel with the 'Troubles'. Other branches of government can with some plausibility claim greater urgency for their concerns than for those of the Environment Service. The army erects look-out stations on hills along the border without regard to visual amenity while security force helicopters cause extensive noise pollution (see the complaints in the submission by the Association of Local Authorities in the Rossi Report, 1990: 107). This is not raised as an issue by any of the principal NGOs. Although these examples are rather far-fetched, the point is that it is difficult to imagine any security policy being overridden by environmental considerations.

Policies carried out in the name of promoting economic activity have not experienced quite such a trouble-free adoption. Although, as I mentioned earlier on, the government has strong reasons for favouring projects likely to create jobs and wealth, a number of proposals have run into strong opposition.

For example, there has been public opposition to various projects aimed at exploiting natural resources, especially when pollution or landscape destruction are likely. Thus, there have been effective residents' actions against plans to prospect for, and presumably then mine, minerals in hills around the historic landmark of the Cave Hill adjacent to Belfast. The Cave

Hill campaign drew support also because of the experience of basalt mining on the nearby Black Mountain, where community campaigners had long been arguing that the mine was being worked beyond the legal limits and without appropriate safeguards to limit noise and particulate pollution. There has been public disquiet over plans to mine lignite in the centre of NI for fuel and to extract gold from the Sperrin Mountains in the north west. Peat extraction has attracted criticism too, though more from environmental organisations than from local residents in most cases. Here there appeared to be clear evidence of governmental double standards since the government was found to be grant-aiding peat-cutting equipment at the same time as the Environment Service was prioritising peat conservation.

In each of these cases the developers sought to use the jobs-versus-the-environment argument. But few jobs were likely to be created from mining or peat extraction in any event, and there were suspicions that any wealth created was unlikely to benefit the majority of the local community. It must also be recognised that these disputes could easily become imbued with sectarian meanings. For example, one proposed site for lignite extraction by open cast mining would have involved the disruption and likely destruction of a nationalist community. Appeals that the mining development should be allowed to go ahead because it would assist the NI economy (or indeed reduce the financial burden on the UK exchequer) could be expected to find little resonance in that community.

Not all of these proposals have been defeated: the gold mining application went to public inquiry in 1993 and the outcome is as yet unknown; lignite mining has not been finally abandoned, it is just not being pursued at present since other fuels are still inexpensive; and residents around the Black Mountain are still very discontent. But communities have demonstrated that they can, at least sometimes, oppose these developments.

Such locally-driven successes have in a sense been complemented by the difficulties faced by the Department of Economic Development (DED) in stimulating investment and relocation to Northern Ireland. The DED's Industrial Development Board's list of overseas investors shows that the majority of foreign companies are involved in engineering and 'traditional' industries, such as clothing, textiles and food processing. There are few cases of firms which are in a position — because they are offering large numbers of jobs while threatening to cause significant pollution – to really bring the issue to a head; in this sense the Northern Irish case contrasts strongly with areas of the Irish Republic such as Cork Harbour, which have attracted multiple investments from chemical and pharmaceutical multinationals. No doubt this is due more to the difficulty of attracting capital investment than to overriding environmental sensitivity on the DED's part.

The one case in the recent past which conspicuously did involve pollution from a large employer centred on Du Pont's plant at Maydown in Londonderry. Du Pont is a major employer in Derry and had trumpeted its high environmental standards. It was, for example, one of the backers for an environmental investigation unit at Queen's University Belfast. But in January 1991 it announced plans to build an incinerator on its site which would burn both waste from its fibre plant and other toxic waste brought to the plant.

The context for this decision was of critical importance. During the 1980s officials in the Irish Republic had prevaricated about their policy on toxic waste disposal. Several plans were brought forward but each was defeated, either by concerted local opposition (typically supported by local members of parliament (TDs) of the ruling and opposition parties, such is the clientalistic condition of electoral competition in the Republic of Ireland) or by a change of government; elections happened frequently during the late 1970s and 1980s. The waste was accumulating, no disposal site was in prospect and communities throughout the country had set their hearts against it. With an incinerator already having been rejected by several communities, no new community was now going to accept one. Suddenly, Derry appeared to be the answer. The Irish Republic's waste could be burnt without it happening in any TD's constituency. However, the plans won no support amongst politicians in Derry either. Nationalists favoured solidarity with the South, reckoning that their tolerance of pollution should be no different from the Republic's. Unionists were appalled at the idea of importing Fenian waste. Community opposition swelled quickly, encouraged by seasoned local community campaigners from formerly targeted sites in the South and by Greenpeace Ireland. A year later the plan was abandoned, though it has never become clear whether the scheme was dropped because of public opposition or because the extent of grant aid for the project was insufficient to make it commercially attractive. In the meantime Du Pont had further weakened their environmental credentials within Derry by causing a large spill of toxic material (chloro-butadine) into the River Foyle in October 1992, even though the Director of Public Prosecutions decided that there were insufficient grounds for a prosecution over the incident.

It was in this policy context of few new large firms that Robert Atkins, then minister for environment and economic development, launched a 'Growing a Green Economy' strategy in March 1993. The claim of his document was that environmental and economic development objectives are more than compatible, they are mutually supportive. He argued that it is possible to 'maintain a growing economy without detriment to the natural and built heritage' (DoENI/DED, 1993: 4).

There are, of course, a number of areas where this argument makes some sense in virtually any economy. For example, companies which use fuel sparingly reduce pollution and cut costs. There are also new economic opportunities for ecological services and equipment which local firms could well be stimulated to exploit. In this spirit the Department of Agriculture NI, accepting that there is buoyant demand for organic food-stuffs, in 1993 proposed assistance for organic farming. Also, owing to the lack of smoke-stack industries, local agricultural and maricultural products benefit from a general green image. Conspicuously Atkins's document focuses on the ways that environmental quality can serve the economy, not the other way round.

So far so good. But there are two points on which his claims were criticised. First, since – as we have seen – environmental law tends to lag in NI, there is little pressure on firms to be at the environmental forefront in their processes or products. If high environmental standards are good for the economy one might think that would be a reason for accelerating the handling of Northern Irish environmental legislation. Second, a lot of industry and agriculture will remain unaffected by the minister's suggestions. To achieve real greening in these cases demands special incentives, tougher legislation or innovative tax arrangements which favour clean business. If he wanted to make economic development and environmental protection mutually supportive in NI in a special way, the minister would have needed to introduce more radically innovative policies. As they stand, his proposals are unlikely to have a rapid or far-reaching impact. In any event, Atkins's successor, an accountant who boasts of no environmental or conservation interests in *Dod's Parliamentary Companion*, does not seem well placed to lead with environmental innovations.

Finally in this section, there are two additional points both of which indicate that environmental performance has to be assessed more broadly than just in terms of explicit environmental policy for NI. The first returns us to the issue of urban air pollution. Earlier I noted that air pollution problems persist in Belfast, Newry and Derry (see DoE NI, 1993: 34). As dirty house coal is generally cheaper than alternative fuels (there are currently no supplies of natural gas), smoky coal is commonly burnt wherever it is permitted, and not infrequently even where it is not. Merely substituting smokeless fuels would raise the cost to the consumer (without even necessarily reducing the *sulphur* content: House of Commons Environment Committee, 1993: 15) while replacement heating systems impose large demands for capital. Such environmental reforms threaten to bear most heavily on the poorer people in NI. Conversion grants are available but these require a lot of administrative work and are a draw on the public purse; they have been distributed in phases over several years. According to officials in the DoE NI, it now appears that gas will be piped in, helping NI to meet its sulphur dioxide reduction targets and assisting in reducing greenhouse gas emissions. Gas may even be made available as a primary fuel for consumers in the Greater Belfast area (House of Commons Environment Committee, 1993: 15).

This example demonstrates both that, on the small-scale local level, the details of environmental policy are inseparable from social policies and that, at the other end of the scale, the broad determinants of environmental performance are commonly set outside NI. Under questioning by the Environment Committee the minister denied that there were plans to pipe gas to Northern Ireland (1990: 96); three years later the policy has been reversed. There appears to be little Northern Irish control over such issues and no practical scope for a Northern Irish policy on acid emissions or greenhouse warming.

After briefly touching on energy policy, the other issue which should be mentioned is transport. This topic barely featured in the Rossi Report or in the *Environment Service Report 1991–93,* yet it is important for air quality in Northern Ireland. In particular, given the relatively low levels of industrialisation, the practical restrictions on changing domestic fuels and the fact that policy towards air pollution from power stations is largely out of the hands of officials in Northern Ireland, this is one of the few areas where action on air pollution and other environmental hazards might be originated locally. However, there is virtually no environmental interest in transport within the DoE NI.

NGOS AND ENVIRONMENTAL POLICY IN NORTHERN IRELAND

Environmental NGOs are acknowledged to be an important influence on policy and practice in Great Britain. They play a key role in NI as well, as acknowledged in the Rossi Report (1990: 87), although they operate in significantly different ways from the rest of the UK. Briefly, there are three major differences. First, the level of public support for NGOs of all sorts tends to be lower than in Great Britain. The RSPB, for example, has around 5000 members; the Ulster Wildlife Trust (or UWT, the local affiliate of the Royal Society for Nature Conservation, or RSNC) around 1000; and Friends of the Earth well under that figure. The green movement has fewer supporters as a proportion of the population; according to the 1991–92 Social Attitudes survey, Northern Irish people are about half as likely to be members of environmental groups as are the inhabitants of the mainland (Stringer, 1992: 26). Northern Irish respondents also expressed less concerned about environmental issues than did their counterparts in Great Britain (1992: 31–2).

The overall profile of the NGO sector also differs from the situation in Great Britain. As indicated above, wildlife conservation interests have until recently been predominant in the NGO sector and this was reflected also in the fact that Friends of the Earth had no staff in NI until 1992 (it now has one person) while Greenpeace had (and has) no one. Transport 2000 is planning to establish itself in 1994. The 'Link' organisation, NIEL (the successor to NIEG, described above), is composed more of conservation and heritage protection groups than of environmental campaign organisations, and its only major report to date (Milton 1990) is concerned principally with countryside and wildlife issues, as its title (*Our Countryside Our Concern*) indicates.

Third, these differences are magnified by the way that governmental support for NGOs tends to operate in NI. DoE NI is willing to grant aid NGOs, including the UWT, Conservation Volunteers and NIEL itself, meeting a high proportion of these groups' core salary bills. In part it is willing to do this because these organisations help to fulfil the mission of the Environment Service, managing the countryside, maintaining reserves, disseminating environmental information and performing advisory services. They have also benefited from a range of job-creation schemes which allows the UWT, for example, to have around 40 employees working on practical projects paid for from a DED budget. Conservation groups can also benefit from EC and other international funds to support wildlife centres and similar ventures. In this way, official (specifically governmental) funding helps to determine the shape of the environmental lobby, and the degree of governmental largesse limits the extent to which some NGOs feel they can antagonise the government.

Furthermore, there is something of a 'revolving door' between the DoE NI and elements (notably the wildlife conservation aspects) of the voluntary sector. This was made clear in Peter Bottomley's replies to the Environment Committee (1990: 85) when, as evidence of the high calibre of DoE NI staff, he cited the case of the former 'Director of Environmental Protection [who] was recruited by one of the major voluntary bodies'. In fact the official in question became Regional Director of the National Trust. Other senior officials have served the UWT and NIEL either as paid officials or as volunteer members of their boards of directors. This undoubtedly lends these organisations significant credibility within the DoE NI itself but threatens to make the organisations too 'understanding' of the limits within which the DoE NI operates.

CONCLUSION

In summary, the pervasiveness of government in NI – as an employer, as a promoter of economic development, as environmental watchdog and as a funder of conservation activities – lends environmental politics a unique and complex quality. While, in principle, having the state as the major player in economic development might be thought favourable to environmental concerns – the state is, after all, more likely to take a long-term view than are private developers – security and economic development seem bound to continue to be the overriding priorities. Environmental policy looks set to remain a poor third and to be essentially derivative of policy in England and Wales. The influence exerted by the visiting Environment Committee of the House of Commons, in a position to examine the DoE NI in a way no other organisation had been able to before, indicates the derivative status of Northern Irish policy and the limited nature of the checks to which it is customarily exposed (see the minister's comments in House of Commons Environment Committee, 1993: 22).

Many of the major issues of environmental policy relevant to NI (greenhouse gas targets, acid emission targets, the availability of natural gas) are not in fact determined in the region, and this position will not change in the 1990s. Still, significant reforms have been introduced affecting many of the areas where government does exercise an influence, resulting in a consolidated Environment Service and a degree of broad monitoring and policy review from the CNCC. However, the minister is still divided between economic development and environmental protection objectives and many commentators do not see those as truly compatible. Moreover, in those areas where there could be a distinctive local policy stance, on transport, for example, and on wind power (prior to privatisation) there has not been evidence of governmental initiative.

Finally, government also shapes the profile of environmental NGOs in Northern Ireland and hence indirectly the character of public action. While there are a number of single-issue, community protest groups (some of which have recorded notable successes) the groups which are in a position to fashion a more comprehensive environmentalist outlook are strongly affected by the state's policies. The more traditional conservation organisations – groups which enjoy good relations with government and which are seen to lobby 'reasonably' – are handed an initial advantage. Whether they will take the opportunity to become a voice for broad environmental concerns or whether that will be left to emerging groups such as Friends of the Earth (still less than a year old in Belfast) has yet to be established.

Note

1.   I should like to thank colleagues present at the meeting in Newcastle in December 1993 and Kay Milton, who was not able to attend that meeting, for their extensive comments on earlier versions of this paper; Kay's insight into the 'independence' of independent agencies was of particular benefit to me.

References

Baker, S. Milton K. and Yearley, S. (eds) (1994), *Protecting the Periphery* (London: Frank Cass).

Bradley, J. F. (1990), 'The Irish Economies: Some Comparisons and Contrasts', in R. I. D. Harris, W. Jefferson and J. E. Spencer (eds), *The Northern Ireland Economy: A Comparative Study in the Economic Development of a Peripheral Region* (London: Longman).

CNCC (1991), *First Report 1989–1990* (Belfast: CNCC).

Dod's Parliamentary Companion (1993) (Hurst Green: Dod's Parliamentary Companion).

DoE NI (1991), *The Government's Response to the First Report from the House of Commons Select Committee on the Environment* (London: HMSO).

DoE NI (1993), *Environment Service Report 1991–93* (Belfast: DoE NI).

DoE NI/DED (1993), *Growing a Green Economy* (Belfast: DoE NI/DED).

Elliott, S. and Wilford, R. A. (1987), 'Administration', in R. H. Buchanan and B. M. Walker (eds), *Province, City and People: Belfast and its Region* (Antrim: Greystone), 285–304.

Harris, R. I. D. (1990), 'Manufacturing Industry', in R. I. D. Harris, C. W. Jefferson and J. E. Spencer (eds), *The Northern Ireland Economy: A Comparative Study in the Economic Development of a Peripheral Region* (London: Longman).

House of Commons Environment Committee (Rossi Report) (1990), *Environmental Issues in Northern Ireland* (London: HMSO).

House of Commons Environment Committee (1993), *Environmental Issues in Northern Ireland: Minutes of Evidence* (London: HMSO.)

Jefferson, C. W. (1987), 'Economy and Employment', in R. H. Buchanan and B. M. Walker (eds), *Province, City and People: Belfast and its Region* (Antrim: Greystone) 191–214.

Milton, K. (1990), *Our Countryside Our Concern: The Policy and Practice of Conservation in Northern Ireland* (Belfast: NIEL).

Richardson, G., Ogus, A. and Burrows, P. (1982), *Policing Pollution, a Study of Regulation and Enforcement* (Oxford: Clarendon Press).

Stringer, P. (1992), 'Environmental Concern', in P. Stringer and G. Robinson (eds), *Social Attitudes in Northern Ireland: The Second Report 1991–1992* (Belfast: Blackstaff), 18–38.

Yearley, S. and Milton, K. (1990), 'Environmentalism and Direct Rule: The Politics and Ethos of Environmental Groups in Northern Ireland', *Built Environment*, 16, 192–202.

7 The Politics of Mutual Attraction? UK Local Authorities and the Europeanisation of Environmental Policy

Stephen Ward

INTRODUCTION

In the UK, local government has traditionally had a strong role in implementing environmental policy and over the course of the last decade it has been one of the most active governmental institutions in pursuing progressive environmental policies (Ward, 1993). At the same time the increasing significance of the EC/EU in the environmental policy field, particularly since the Single European Act (SEA), has considerable implications for the subnational sector. It has been argued that the EC has acted in the past as a centralising force in environmental policy relations, indirectly reducing some of local authorities' statutory environmental responsibilities (Haigh, 1986: 205). Yet it is possible in recent years to detect the beginnings of a relationship between the EC/EU and local government in the environmental sector, which some local government actors claim presents authorities with potential opportunities to develop a stronger environmental role.

The arguments outlined below provide an overview of local authority relations with the EU in the environmental policy arena. Such discussions draw partly on inter-governmental relations theories. I will argue that both the EU and local government are to some extent dependent on each other for both practical policy implementation and for wider democratic purposes. However it needs to be remembered that the relationship is still in its infancy and there are significant obstacles, not least the attitude of national government towards a stable long term input for sub national government.

THE INTERGOVERNMENTAL RELATIONS APPROACH

Inter-governmental relations (relations between the EU, national government, and subcentral government) have been classified in terms of three models: hierarchical, consultative, and participative (see below, and Mawson and Gibney, 1985: 136; Rhodes, 1986, 46).

1. The *hierarchical* model describes a top down relationship where there is no significant contact between supranational government (SNG) and subcentral government (SCG). Instead contact is mediated by national government (NG) which passes down policies to SCG and acts as a gatekeeper of information.
2. The *consultative* model indicates a triangular relationship, with SNG involved in consultative policy relationships with both NG and SCG. These consultations are more of a mutual exchange of views. SCG can either lobby through the NG or form its own links with the SNG.
3. The *participative* model describes a complex web of policy networks indicating an even stronger role for SCG. In essence SCG can participate in five different ways:
    (a) SCG lobby NG, which then lobbies SNG.
    (b) SCG lobby SNG directly.
    (c) NG and SCG jointly lobby SNG.
    (d) SCG lobby SNG through SCG associations.
    (e) SCG form their own networks with other local authorities and lobby SNG.

Figure 7.1   Intergovernmental relationship models

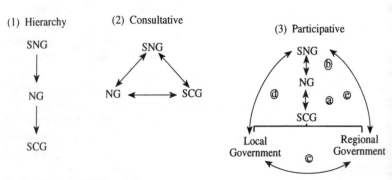

Source: Rhodes (1986, p.46).

Rhodes has argued that the type of intergovernmental relationship is dependent on four resource factors (1986: 47–51).

1. *Legal*: the constitutional position of SCG and the distribution of functions and competences is one important variable. Where the position of SCG is constitutionally entrenched and has strictly guarded functions, its position is likely to be much stronger.
2. *Political*: the strategies and interests of participants at each level are important. Thus a key factor is the desire of SNG and NG to seek alliances with SCG to bolster their own position. At the local level the ability of leaders coherently to represent their region is a determinant of any sucessful strategy.
3. *Organisational*: this refers to the relative availability of expertise, staff and equipment at each level. The extent to which SNG or NG is dependent on SCG for implementation affects the likelihood of SCG being involved in policy making. Conversely, SCG's ability to make use of any specialisation is dependent on its knowledge of European affairs.
4. *Financial*: funding from SNG is an important factor in determining relationships. The existence of funds often encourages participation from both SCG and NG. Each level, but particularly NG, often attempts to control discretion over the distribution of funds.

Rhodes argues that relations generally are in the process of shifting from the hierarchical to the consultative. However, this does not, he argues, indicate the relative influence of various tiers of government. Essentially NGs are still the dominant force because SNG lacks the discretion in relations and SCGs have difficulties in terms of information gathering and, moreover, overcoming the fact that they represent a multiplicity of often divergent interests (1986: 47).

In the local government literature, much stress is placed on the growth in awareness of the EU, the increasing number of links with it, and the development of the concepts of partnership and subsidiarity, particularly in the environmental policy field (Preston, 1991; Association of County Councils – ACC, Association of District Councils – ADC, Association of Metropolitan Authorities – AMA, Local Government Management Board – LGMB, 1992; Roberts, Hart and Thomas, 1993). Environment, it is argued, is a policy area where local authorities are particularly well informed with regard to EC affairs. It should therefore be possible to test whether in the environmental policy sphere inter-governmental relations are actually moving away from the traditional hierachical model towards participative models.

LOCAL GOVERNMENT AND EUROPEAN ENVIRONMENT POLICY: MUTUAL DISCOVERY

In the Treaty of Rome, neither SNG nor environmental policy is mentioned. Although the Community developed both environmental and regional policies, neither of these explicitly recognised the importance of local authorities, and nor was there much attempt to coordinate the two policy fields (Mazey and Richardson, 1993). Until the mid-1980s progress in both fields was disjointed and piecemeal. Thus the signing of the SEA proved to be important in a number of respects. Firstly, it marked a formal recognition of the EC's growing involvement with environmental matters. Article VII specifically provided the legal basis for EC environmental policy and also contained a subsidiarity clause which implied a division of competences between member states and the Community, but did not mention the role of SCG in this respect.

Secondly, the SEA increased the process of recognition of the importance of local authorities in European integration, with the promotion of European regional policy and emphasis on the concept of Europe of the Regions. Traditionally regional policy has been viewed as a vehicle for economic development in the poorer regions of the Community, often leading to significant environmental degradation. However, over the last two years the Community's structural funds have come under pressure from environmental organisations and from a highly critical Court of Auditors' report to incorporate stricter environmental controls (EC Court of Auditors, 1992).

The SEA also incorporated the notion that the environment should be taken into account in all Community policies and therefore implied the integration of environmental and regional policy objectives. This, along with efforts to green the structural funds, marks the beginning of a partial convergence of environmental and regional policy.

The role of local government suggested in the SEA has since been strengthened by the Maastricht Treaty which explicitly outlined an input for local and regional authorities through the formal creation of a Committee of the Regions consisting of elected representatives from local and regional authorities in member states.

The increasing formal recognition has led to emphasis on the twin themes of partnership and subsidiarity. The former indicates a desire for increased inter-governmental cooperation in terms of both policy making and policy implementation; the latter stresses the importance that decisions should 'only be taken at Community level where this would be more effective than action taken by the Member States' (Verhoeve, Bennett and

Wilkinson, 1992: 12). Interestingly whilst formal recognition of local authorities has been a recent development, subsidiarity has always been part of the environmental policy sector although not explicitly set out until the SEA (Haigh, 1990: 10).

The convergence of the general recognition of local authorities and increasing emphasis on environmental matters can be illustrated by a range of policy documents in the environmental area which all emphasise the important role of SNG.[1]

The urban environment Green Paper issued in 1990 proved to be a seminal document. Most environmental legislation until then had been of a sectoral rather than a spatial nature. The Commission's previous incursion into local planning matters had been the introduction of EIAs, which had aroused significant controversy over the legitimacy of the EC to act in such circumstances. Arguably, therefore, the urban Green Paper demanded a new approach that required the support of urban local authorities. Hence the Commission embarked, probably for the first time in the environmental field, on wide scale consultation, through a series of national seminars with local authorities and other interested parties, about the contents of the paper. The paper sparked off a good deal of debate about the role of the local authorities and indeed the Commission in the urban environment. The Green Paper itself noted: 'The primary focus for action to improve the urban environment is clearly the individual city. But achieving major improvements will require action at national and Community level as well, with roles and responsibilities assigned to each within a framework of cooperative partnership' (Commission of the European Communities, 1990: 2). Whilst there was some concern about interference from the Commission in what certain local authorities felt to be their primary concerns, there was a good deal of positive feedback. The real importance of the document was that it brought the two levels of government together discussing policy issues and agendas and demonstrated to the Commission that local authorities had a great deal to offer. Paradoxically despite the claims of interference, it brought DG XI directly closer to a greater number of local authority actors.

The Environmental Fifth Action Programme clearly continues to identify the importance of local authorities, arguing for a relationship based on partnership between all levels of government particularly in terms of implementation of an environmental programme. It states: 'local and regional authorities have a particularly important part to play in ensuring the sustainability of development through exercise of their statutory functions as competent authorities for many of the existing Directives and regulations and in the context of practical application of the principle

of subsidiarity' (Commission of the European Communities, 1992: 26).
Discovery has not been a one way process. Local authorities also have
begun increasingly to make their own policy, organisational and structural
links to the Community. As local authority environmental plan documents
(environmental charters, audits and action plans) have developed they
have begun to incorporate European overtones (Ward, 1993). The local
government associations' guide to environmental practice has a specific
European focus and was part sponsored by DG XI (ACC, ADC, AMA,
LGMB, 1992). The LGMB has also brought out a guide to the Fifth
Action Programme (LGMB, 1993). Similarly work in progress on local
sustainability and developing local governments' follow-up to Agenda 21
has indirectly brought local authorities into further contact with the Fifth
Action Programme and its emphasis on sustainability.

Local government has been eager to develop its own experience of
environmental auditing and link into the EU's eco audit regulations. The
local government associations and the DoE have supported pilot schemes
in a number of authorities seeking to adapt the audit regulation.

EU–Local Government Environment Structures

European environmental policy documents illustrate recognition of the
contribution of the local/regional level, but there has also been a development of organisational structures which have allowed local authorities an
input into EU environment policy making. Two particular European committees have enabled local government representatives to engage in
dialogue with the Commission: the Council for European Municipalities
and Regions' (CEMR) Environment Committee (previously called
the Environment Dialogue Group, or EDG) and the Urban Environment
Committee.

The original EDG was initiated in 1987 by the International Union of
Local Authorities (IULA), with support from DG XI. This grouping consists of local authority representatives from member states and also occasionally representatives from non-EU countries. The British delegation has
always been the largest with around ten to fifteen members, whereas most
other member states have usually sent fewer than five delegates. The dele-
gations have until recently consisted largely of specialists in environmental policy. For the British delegation this has meant mainly officers rather
than elected councillors.

As part of the formalisation of its position, from 1993 the EDG has been
re-organised as the CEMR's Environment Committee. The change
requires that it comprise elected representatives. Thus the British

delegation will now consist of councillors nominated by the local government associations. At the time of writing the new committee is in a state of flux, but it is suggested that the original officer dialogue group will remain to provide specialist back up to its deliberations. However, direct contact with the Commission will now be the new committee's preserve. The changed arrangements have the greatest impact on the British delegation, since for other member states there is not such a sharp division between elected councillors and technical officers as specialist officials are often elected politicians. Previously the EDG has met once a year in full session primarily to discuss the environmental work programme of the Commission for the coming year and environmental issues of particular concern to local authorities.

The value of the EDG for local authorities is that it gives them one of the few early inputs into the Commission's thinking and there is a feeling from the British delegation that their views are listened to more sympathetically than they would be at Whitehall. It also provides an opportunity for local authorities to discover what their European counterparts are involved in environmentally (i.e., it acts as a conduit of good environmental practice).

From the Commission's point of view, the EDG has provided technical advice from ground level implementers of EU directives and it has helped provide a small directorate such as DG XI with an additional base of support for its ideas. One EDG participant suggested that DG XI's openness could be explained partly by the fact that as a relatively small directorate with limited resources it needed to gain friends wherever it could in order to win battles elsewhere within the Commission (interview, 24 August 1993).

Local authority representatives also participate in a technical grouping known as the Urban Environment Expert Group, or Urban Environment Committee (UEG). This committee was set up by DG XI in the aftermath of the urban environment Green Paper to bring in specialist advice and develop a more sophisticated urban policy The group contains six local authority specialists from a variety of urban areas (Bath, Evora, Athens, Aarhuis, Nancy and Stuttgart); three other specialists (an architect, an urbanist, and an ecologist); and twelve civil servants from each of the member states.

The group meets two or three times a year, and has mainly taken its agenda from the Green Paper, though its focus is fairly wide ranging and covers virtually any Commission activity which has implications for the urban environment. Following re-organisation at its recent meeting the main focus of the group is now sustainable development in cities.

EXPLAINING ENVIRONMENTAL CONVERGENCE: MONEY AND
THE NEW LEGITIMACY

The increasing convergence of the environmental thinking of the EU and
local authorities results from a growing realisation of the mutual benefits
of partnership, not only in terms of achieving successful policy implementation, but for wider democratic purposes. The main factors behind this
evolving relationship are as follows.

1   Legitimation

Both the EU and local government are using each other in order to support
and legitimise their own environmental activities. Because environmental
policy is an area which receives wide public support in contrast to some
other EU policy arenas, local authorities can appeal to the EU as a higher
source of authority when faced with resistance or apathy from NG or other
public agencies. Similarly local authorities are also justifying environmental policy activity in European terms, by producing environmental projects
which reflect European thinking. Roberts, Hart and Thomas argue that
where 'local authorities have their own environmental aspirations, it has
often been found that the stance and policies of the Community have provided moral and practical support for these aspirations, generally helped in
attaining legitimacy and sometimes provided support for the higher than
national standards that local authorities were seeking to promote'
(Roberts, Hart and Thomas 1993: 97). This creeping Europeanisation of
their own environmental work is often viewed in a positive light, with
authorities seeing themselves as progressive, forward thinking and at the
forefront of environmental change. UK local authorities clearly believe
that their efforts and aspirations in the environmental field are ideologically compatible with those of the EU.

For its part the EU has learnt to stress the local implications of its environmental policy in order to meet the criterion of subsidiarity and to
attempt to demonstrate that its growing involvement in the environmental
sector does not necessarily involve increased centralisation, despite the
often cross-boundary nature of environmental problems. Indeed in some
ways it is attempting its own version of the much used 'think global, act
local' maxim. By actively courting local authorities the EU can go some
way to overcoming the democratic deficit.

One further aspect of this process of legitimation is that the representatives on local authority-EU environment structures see part of their role as
not only representing local government, but also representing the work

and views of the Commission in local government circles. One British representative on the UEG argues, 'the work of the group provides, in my opinion, some form of legitimation process for the work of the Commission and DG XI in particular' (Fudge, 1992: 6).

2 By-Passing National Government

Where necessary both the EU and local authorities have recognised the value of by-passing difficult NGs. Local authorities generally regard the EU as more environmentally sympathetic than NG. DG XI is seen as offering a more receptive arena for local authorities to discuss environmental problems, as one local government observer of the EDG noted:

> Certainly DGXI is very keen to involve local authorities in the consultation process. They actually have a lot of respect for local authorities and I think that local authorities who aren't used to that from central government here, warm to the fact that they are being consulted and put a lot into making sure the Commission gets some feedback (interview, 28 July 1993).

There is a second aspect to by-passing which has resulted from the changed legal position of the subcentral unit in Europe where the EU has increasingly become the senior legal tier. Local authorities can now appeal over the head of NGs to the EU. Hence local authorities can potentially play what might be described as a whistle blowing role, in that they can monitor NG's implementation of directives in their locality. For example, Lancashire County Council took the British Government to the European court over sewage problems on the North-West's beaches and breaches of the Bathing Water Directive (Stoker and Young, 1993: 159).

3 Implementation and Monitoring

There has been an increased debate focusing on improving and monitoring the implementation of environmental directives. Since local authorities are one of the key organisations in administering environmental directives and regulations in Britain, they have taken the opportunity to complain of their relative lack of an early input into the EU policy process and to argue that incorporation of their specialist and local knowledge could smooth the implementation of directives and reduce the failure rate.

The EU itself has also begun to recognise the value of consolidating and improving existing directives and regulation, rather than simply seeing

environmental success in terms of introducing more and more new legislation. The Fifth Action Programme devotes a whole chapter to the problem
of implementation and contains specific proposals for improvement by
creating implementation networks which include local authorities.

4 Funding

Often the attraction of Europe is stimulated by the possibilities of winning
extra funding for projects. In the current harsh financial climate 'Europe'
is seen as a pot of gold. Many authorities have become adept at seeking
funding through the European Regional Development Fund (ERDF) and
have extended this expertise into the environmental LIFE (EU Financial
Instrument for the Environment) programme. In order to be able to implement principles drawn up in environmental charters/audits many authorities see the EU as a funding lifeline, or as a last resort once local and
national government funding sources have been exhausted. The popularity
of LIFE is further enhanced because it is open to all local authorities.
Unlike regional funds there are few restrictions regarding who can apply,
though there is much less cash and DG XI officials have repeatedly
warned against seeing the EU as a pot of gold (London Boroughs
Association – LBA, 1992: 10).

5 Euro links

A developing trend in the course of the last decade has been the growth of
European and international networks, and environment policy is a particularly fertile area for cross-national environmental links. Environmental
networks have a number of functions. Firstly they are about exchanges of
good environmental practice and stimulating innovative environmental
projects. Secondly they allow strategic thinking about cross-boundary
environmental problems not simply between national or international governmental levels. Thirdly networkers have recognised the benefits of
European links in terms of increasing their environmental profile before
the Commission, in order to assist in lobbying and competing for funding.
Local authorities have realised that chances of successful access and
funding from the Commission are boosted by a cross-national dimension
(Parkinson, 1992; Robson, 1992). Thus networks have both stimulated an
interest in the EU and been supported by the EU.
    The EU's support for networks is for both practical and ideological
reasons. Ideologically the EU is keen to foster cross-border links as a

means of breaking down national boundaries and encouraging groups to think European. In practical terms, if networks are coordinated and present unified representative information or arguments, then the Commission has to spend less time analysing information from twelve different member states each with its own organisations, culture and politics.

6   Professional Values

The professional groups primarily dealing with the bulk of European environment legislation are planners and environmental health officers. Rather than resenting the increasing role of the EU as a threat to their professional values, they tend to focus on its potential benefits. Their marginalisation during the last decade in terms of national politics has perhaps made them more receptive to the EU style of environmental regulation. They tend to use European environmental issues as an opportunity to raise the profile and importance of their professions. One leading policy officer from the local government associations said of Environmental Health Officers (EHOs),

> [they] are regulators and enforcers by the nature of their work and therefore they don't have a problem with European directives ... They will sometimes have criticisms about exactly how it's done, or the lack of resources, or the lack of time allowed to implement, but I wouldn't say in general that there is a mood in local government that's opposed to what they see as a European project to toughen up control of the environment and health, they favour it, (interview, 26 July 1993).

IMPEDIMENTS TO A SUCCESSFUL ENVIRONMENTAL MARRIAGE

There is a danger of overstating the areas of convergence in the environmental policy sector between local authorities and the EU. There are still a number of major obstacles which are preventing the formation of a fully functioning partnership in the environmental sector and weakening local authorities' involvement in developing policy networks. Many of the problems are internal organisational or structural ones for local government in Britain. However, perhaps the most serious obstacles are exogenous ones, out of local authorities' control, and particularly the attitude of NG.

1 Relations With National Government

Despite the increasing institutionalisation of local authority links with DG XI there is still a widespread belief that local authorities are shut out of the early stages of European policy discussions. Until recently, the DoE view appeared to be that local government had no need to be involved in policy negotiation until European legislation was transposed into domestic legislation. When giving evidence to a House of Lords Select Committee investigating implementation of EU environmental legislation, the AMA complained on a number of occasions about their lack of involvement in European policy making:

> No formal mechanism currently exists to allow important dialogue to occur ... This can prove extremely detrimental leaving local authorities almost completely reliant on central government for information which at times can be severely limited ... Although consultation does sometimes occur when local authorities are directly involved in implementation, it is very much the decision of the department and not a statutory requirement. This would appear to be a curiously British phenomenon, leaving local authorities feeling alienated from the decision making process. (House of Lords Select Committee of the European Communities, 1992: 91)

Whilst the DoE are content to see local authorities actively engaged in funding activities – the local authority associations noted that the DoE had been very supportive in bids for LIFE funding – NG is clearly more circumspect about allowing local authorities any influence on policy which may disturb its own position. Encouraging local authorities to win funding represents little threat to the government's stance and, with tight controls on overall spending levels, it may well reduce the PSBR. It may also be thought that the more time local authorities spend searching for funding the less time there is for policy inputting.

Those authorities which take a reactive policy stance are clearly quite dependent on the DoE for European environmental information, yet there is some feeling that the DoE fails to provide enough relevant information, and possibly even filters the information it provides. The Commission may have initiated a policy of access to environmental information, but unless authorities go direct to EU, local government participants complain, it is difficult to get hold of information (Morphet, 1992b: 5).

2 Fragmentation of Opinion

The alternative danger is that as more and more authorities take an individual path to Brussels the potential for a coherent unified local authority standpoint is reduced. This fragmentation reflects in part the inability of the local authority associations adequately to represent local authorities at the European level. This produces problems in turn for the Commission, which prefers to deal with groups which provide a coordinated, authoritative and representative argument from their particular sector. As one leading figure in the EDG commented:

> Our problem is that we've got the Institute of Environmental Health Officers (IEHO) over there, Institute of Trading Standards over there, the Institute of Waste Management over there. We've got Birmingham, Bournemouth, some quite unimportant place in the west country all opening offices over there and all waving their own particular flag ... I think we've got to be very careful when we go over there that we're seen to be saying the same thing, because when you don't say the same thing, they either play us off one against the other or ignore us because they can't understand us. (interview, 7 September 1993).

Hence by-passing can lead to fragmentation of coherent policy argument and could eventually lead to an overloaded policy process and even restriction on local authority input.

3 Information Difficulties

If coordinated inputting into environmental policy in Europe is difficult for local authorities, then widespread organised dissemination of the information gained from contact with Brussels is even more problematic. There are few systematic routes of information gathering or dissemination. For example, whilst the EDG is recognised as being a valuable source of dialogue with the Commission, there is concern that very few authorities are actually aware of its presence or its agenda. Mechanisms for disseminating the information appear to rely largely on *ad hoc* informal structures or individual representatives in the local government world. The main method of passing reports of EDG work down to authorities is through the associations and in particular the LGIB (Local Government International Bureau). Whilst a steering committee of British representatives produces reports, and details of the conference are reported in the

LGIB's journal EIS (*European Information Service*), there is little evidence that this information ever reaches the majority of officers at authority level. One director of environmental services active in European environmental affairs commented, 'I'm not sure people see EDG stuff, and there's a whole week's meeting. I've never seen reports of what the EDG are doing. Unless you look for those things, I'm not sure you'll easily pick them up' (interview 26 September 1993). This last point indicates a further problem of information dissemination, that of internal organisation. EU environment information crosses traditional departmental boundaries and therefore is not easily disseminated. There is a problem of who to direct information at. An obvious coordinating point is an authority's European officer, but this officer's role tends to be orientated towards funding issues, rather than policy inputting, or policy dissemination.

4   Elite Divergence

Due to the relatively *ad hoc* nature of responses to EU environmental initiatives and the dependency on individual officers, three categories of response are developing: a small group of pioneering environmental authorities which are leading the way by developing an expertise in both policy and funding areas; a larger second tier encompassing the bulk of authorities who are mainly concerned with funding opportunities; and a third group which lags behind on both policy and funding. Authorities who were acknowledged as pioneers in environmental policy (Kirklees, Lancashire, Leicester) have developed a better understanding of the increasing importance of supranational institutions in environmental policy formation, and such authorities have attempted to globalise their own agenda. Their discovery of Europe may be partly because these authorities developed environmental policies earlier than others so have had more time to discover European environment policy and policy makers.

   Furthermore it is noticeable that because of the inability of current structures to disseminate information efficiently, at present local authority policies and structures regarding the EU environment arena are being determined by a relatively small network of highly informed individuals, dominated by officers rather than members. Many authorities are failing to input into the current system. Members in particular are taking a relatively reactive approach towards European environment. There appears here to be a double disadvantage, with environmental issues sometimes perceived as being technical issues for officers, whilst European issues are often viewed as being relatively distant from the day-to-day mundane and localised environmental concerns of councillors. Hence the links between domestic legislation and EU directives or programmes are often poorly understood. As one Councillor commented in relation to the Fifth Action Programme:

> [in the average local authority] I suspect the councillors think it not very important. They wouldn't notice it a lot. I suspect if it was called the Fifth Action Programme for X District, we might do something. Again I rather think that the designation of programmes like that need people to interpret them. It's not enough to publish bits of paper. (interview, 7 October 1993).

The councillors' difficulties also arise because there are divisions within the major parties on European issues which has led to a lack of clear direction for party politicians, and neither is there likely to be much enthusiasm from the councillors' own electorate to push them towards European matters. Local politics is not perceived to be about a distant Brussels. Much therefore is still dependent on the attitude of the individual councillor and individual authority towards Europe. The study of local government environmental policy field would seem to bear out George's conclusion that Britain has adapted more at the administrative level than at the political level (George, 1991: 203).

5 The Domestic Local Agenda

It is sometimes argued that Britain is at a disadvantage in responding to the EU because, unlike most member states, it does not possess a regional tier of government (the level of authority with which the EU particularly wishes to develop relations). The current review of local government structure has implications for the debate concerning regional government. The review seems set to take us further away from a European model, by introducing a rather confusing mixed system of local structures across the country. County councils, which in England and Wales see themselves as the regional tier of government, are in some areas under threat of abolition. Despite its apparent distance from European environmental policy, this domestic restructuring debate has some important implications for environmental policy. The Europe 2000 initiative points towards harmonisation of the European planning system and a Europe of the Regions, leading to an increased role for local authorities to cooperate to produce regional plans. Current re-organisation seems to weaken the influence of authorities in this process (Morphet, 1992b, 17).

In broader terms the review is also proving a distraction from the European agenda. As authorities concentrate on domestic restructuring, they are busy defending their own internal interests and for some this is a fundamental issue of survival. To an extent this means competing with one another to avoid abolition, a situation which is hardly likely to engender a common local authority voice.

FROM REACTION TO PROACTION? NEW INTER-GOVERNMENTAL RELATIONS IN THE ENVIRONMENT

In analysing how the growing convergence of local authorities and the EU in the environmental sector has affected inter-governmental relations, it is currently difficult to identify a pattern which definitively characterises relations between the three levels. It is possible loosely to apply all three of Rhodes's models (hierachical, consultative, and participative) in varying circumstances. There is a diversity of approaches from authorities in their relations with the EU. A small but increasing number of authorities have an essentially proactive relationship with Europe in the environmental sector. Figure 7.2 illustrates ten potential routes by which proactive authorites are inputting into European environment policy, including direct contacts with DG XI, local and European networks, the use of MEPs and the professional bodies as channels of influence, as well as the more traditional routes of operating through government departments and the local authority associations.

Since an increasing number of authorities now have direct contact with Brussels, and because of the apparent convergence of environmental aspirations there is a danger of overemphasising the proactive model. For many smaller district authorities the overall tone of inter-governmental relationships is still hierarchical. Many local authorities still have little direct contact, apart from funding inquiries and are largely reactive in terms of EU environment policy. Consequently their level of knowledge of EU environment policy and EU environmental policy procedure is limited. Reactive authorities are largely reliant on two indirect routes through the DoE and the local associations for their information, and their active input back through these organisations is very limited (see Figure 7.3).

The level of activity in European environmental policy matters is dependent on five main factors.

1.   Individual policy entrepreneurs who are active, knowledgeable and have links with Europe concerning EU environmental policy trans-

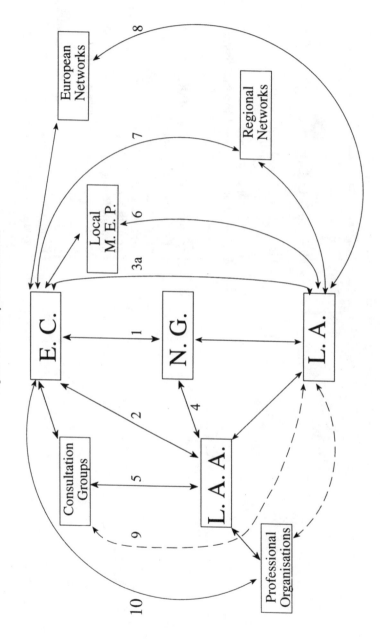

Figure 7.2 The pro active model

Figure 7.3 The relative model

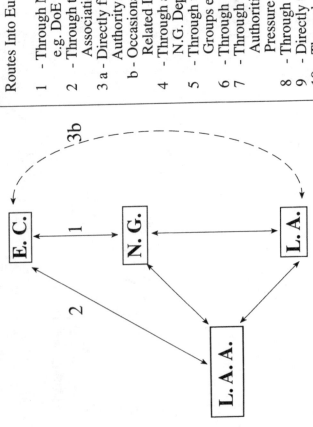

Routes Into Europe

1 - Through N.G. Departments
    e.g. DoE
2 - Through the Local Authority
    Associations (L.A.A.)
3 a - Directly from the Individual
    Authority on Policy and Finance Issues
  b - Occasional Contact on Finance
    Related Issues
4 - Through a Combination of the
    N.G. Department and the L.A.As
5 - Through the L.A.As to Consultation
    Groups eg. the E.D.G
6 - Through a Local M.E.P.
7 - Through Regional Networks with other UK
    Authorities/Local Business/Local
    Pressure Groups
8 - Through European/International Networks
9 - Directly to Consultation Groups
10 - Through the Professional bodies
    e.g. the IEHO

fer their knowledge to others within the authority, creating an internal European network. These actors are normally senior officers.

2. Where authorities have a general European strategy and links to Europe in other policy areas, knowledge appears to spill over from one policy arena to another in European terms.

3. Those local authorities which have pursued EU environmental funding have tended to expand this interest into policy matters. The bidding process itself increases awareness of EU policy, since authorities generally develop a background knowledge of the policy area in which they are bidding for funds.

4. As was noted above, authorities who have been the most environmentally active over longer periods have discovered Europe and are using it proactively.

5. Generally, the upper tier of local authorities, (i.e., regional authorities in Scotland, metropolitan and county councils in England and Wales, along with the larger urban district councils) tend to be the most proactive. These are authorities which have resources and the administrative capacity to make a strategy feasible, as well as having statutory strategic functions which make the Community appear more relevant to the authorities' needs (Martin, 1993).

Relationships between local authorities and Europe are also somewhat variable depending on which part of NG is affected. As with local authorities some departments and some divisions within departments are more proactive regarding Europe than others, and some are more sympathetic towards local authority views. Local government participants in environment policy have noted that there are differences in the attitude of the DoE, and have suggested that internally government departments may not always have a clear voice. The local authority association officers drew a distinction between civil servants in the Environmental Protection division of the DoE and other sections. Their judgement was that the Environmental Protection division was restrained in its relations with local government by its inability to win arguments not only within the DoE, but also with other NG departments.

Despite the moves towards proaction among authorities, therefore, it would appear that local authorities have difficulty deciding whether to avoid the centre and act as a pressure group, with the resultant danger that this further weakens their position with NG. It is important to remember that NG is still a key player in local authority environmental policy despite the increased role of the EU.

CONCLUSION

Local government has the dilemma of wishing to be seen as a statutory consultee with a privileged position and yet being forced to compete as a pressure group. In both roles, local authorities are campaigning from positions of weakness. Theoretically in environmental terms they have, as noted above, advantages which should make them a powerful player in terms of lobbying DG XI, but as a collective organisation they have yet to come to terms with such a lobbying role. One councillor involved in European lobbying both for an environmental pressure group (the RSPB) and for local authorities collectively, argued:

> The difference I notice is that within the RSPB we have a very clear mission. We know what we want to do. We analyse the problems we are dealing with quite clearly ... There is a certain advantage in having a cogent brief, whereas local government comes with a multiplicity of objectives and one suspects that sometimes what local authorities end up doing is arguing between the different objectives rather than trying to resolve them (interview, 7 October 1993).

The difficulty in characterising relationships is further hampered by the current state of flux in terms of both structural changes at EU level and the current restructuring of local government. At present no one is entirely certain how structural changes such as the Committee of the Regions, the proposed environment forum and the new CEMR environment committee will work. Certainly much hope is being invested in the new structures from the likely local government participants. The Committee of the Regions, in particular, has the potential to strengthen local government's position and weaken the hierarchical system as it has statutory consultation rights for local and regional authorities, though at the time of writing there is much doubt as to how successfully it will operate, and whether it will be able to achieve actual influence on policy due to its status.

Whilst NG might be keen to guard the hierarchical style, dissatisfaction and pressure from within SCG and from the SNG end is beginning to erode and fragment the model. However, the corresponding increase in the amount of direct contact with the EU on environmental policy has not necessarily yet increased local influence in any measurable policy terms. The relationship is still in essence about information exchanges and attempts to increase mutual understanding. The ability of local authorities to shape the agenda is distinctly marginal and policy inputs are somewhat ad hoc. Even where consultation operates, it is on a bilateral rather than trilateral basis: that is, discussions between local authorities and the EU (the EDG), NGs and SCGs (the Central–Local environmental forum), or national and supranational institutions.

Note

1. See, for example, the Fifth Action Programme, the urban environment Green Paper, or the White Paper on sustainable mobility.

References

ACC, ADC, AMA, LGMB (1992), *Environmental Practice in Local Government*, 2nd Edn (Luton: LGMB).

Commission of the European Communities (1990), *Green Paper on the Urban Environment* (Brussels: Commission of the European Communities).

Commission of the European Communities (1992), *Towards Sustainability: The Fifth Environmental Action Programme* (Brussels: Commission of the European Communities).

EC Court of Auditors (1992), *Special Report No. 3/92*, Environment.

Fudge, C. (1992), 'The Urban Environment Expert Group', *EIS*, 130, 5–7.

George, S. (1991), *The Politics of Semi Detachment: Britain and the EC* (Oxford: Oxford University Press).

Haigh, N. (1986), 'Devolved Responsibility and Centralization: Effects of EEC Environmental Policy', *Public Administration*, 64, 197–207.

Haigh, N. (1990), *EEC Environmental Policy and Britain*, 2nd Edn (Harlow: Longman).

House of Lords' Select Committee of the European Communities (1992), *The Implementation of EC Environmental Legislation* (London: HMSO).

LGMB (1993), *The EC's Fifth Action Programme on the Environment: A Guide for Local Authorities* (Luton: LGMB).

LBA London Research Centre, Association of London Authorities (1992), *Europe and the Urban Environment* (London: LBA).

Martin, S. (1993), 'The Europeanisation of Local Authorities: Challenges for Rural Areas', *Journal of Rural Studies* 9, 2, 153–61.

Mawson, J. and Gibney, J. (1985), 'English and Welsh Local Government and the European Community', in M. Keating and B. Jones (eds), *Regions in the European Community* (Oxford: Oxford University Press), 133–59.

Mazey, S. and Richardson, J. (1993), 'Policy Co-ordination in Brussels: Environment and Regional Policy', *EPPI Occasional Paper No.93/5* (University of Warwick: Coventry).

Morphet, J. (1992a), 'Eurospeak Staging Post', *Planning*, 974, 26 June 1992.

Morphet, J. (1992b), 'Mandarins lay Strategy to Seize the Regional Ground', *Planning*, 989, 9 October 1992.

Parkinson, M. (1992), 'City Links', *Town and Country Planning*, September 1992, 235–6.

Preston, J. (1991), 'Local Government and the European Community', in S. George (ed.), *The Politics of Semi Detachment: Britain and the EC* (Oxford: Oxford University Press), 104–18.

Rhodes, R. A. W. (1986), *European Policy Making, Implementation and Sub Central Governments: A Survey* (Maastricht: European Institute of Public Administration).

Roberts, P., Hart, T. and Thomas, K. (1993), *Europe: A Handbook for Local Authorities* (Manchester: Centre for Local Economic Strategies).

Robson, B. (1992), 'Competing and Collaborating through Urban Networks', *Town and Country Planning,* September 1992, 236–8.

Stoker, G. and Young, S. (1993), *Cities in the 1990s* (London: Longman).

Verhoeve, B. H., Bennett, G. and Wilkinson, D. (1992), *Maastricht and the Environment* (Arnhem: IEEP).

Ward, S. (1993), 'Thinking Global, Acting Local: UK Local Authorities and their Environmental Plans', *Environmental Politics*, 2, 3, 453–78.

8 The UK and Global Warming Policy

David Maddison and David Pearce[1]

INTRODUCTION

The UK's commitment to the Framework Convention on Climate Change obliges the country to return its emission of greenhouse gases (GHGs) to 1990 levels by the year 2000. On the basis of current trends this requires a reduction of 10 million tonnes of carbon relative to baseline emissions. Earlier the Government had, in its White Paper on the environment, unambiguously endorsed the use of market based instruments to achieve this objective. But in the event the measures which were chosen included imposing VAT on domestic energy, increasing road fuel duty as well as information campaigns and moral persuasion.

Imposing VAT on domestic energy made the government deeply unpopular because of the adverse distributional implications. To some extent, however, this aspect of the tax has been overcome by recent adjustments to social benefits. Whilst increasing the duty on road fuel is not the ideal measure to tackle Britain's transport problems it may, in the interim, reduce the high social costs associated with road transport and, unlike VAT on domestic energy, the burden of increased road fuel duty falls more heavily on higher income households.

The Government has stubbornly resisted the imposition of a European carbon energy tax, ostensibly on the grounds of competitiveness. In reality the British Government's opposition to the tax had more to do with a desire to retain, and to be seen to retain, control over domestic taxation matters. To a lesser extent the Government also feared the political consequences of closing more coal mines so soon after the pits review. Nevertheless a carbon tax will undoubtedly be required if scientific evidence points to the need for greater cutbacks in GHG emissions and may, contrary to expectation, increase both output and employment.

World wide efforts to reduce GHG emissions now have a greater salience. Although some scientists believe that it is still too early unequivocally to attribute recent trends in global temperatures to the existence of an enhanced greenhouse effect, statistical evidence continues to link GHG concentrations and global temperature anomalies (see Tol and de Vos,

1994). Sea levels, too, are rising in line with calculations made by scientists (Tol, 1994) and that the mean temperature of Central England is changing is almost undeniable (Maddison, 1994). But whereas evidence pointing to the existence of an enhanced greenhouse effect continues to accumulate, the impact that climate change will have on ecosystems remains largely a matter for conjecture.

In June 1992 the Prime Minister, John Major, signed the Framework Convention on Climate Change at the United Nations Conference on Environment and Development (UNCED) in Rio. This Convention has now been ratified by Parliament and will shortly enter into force. The Convention compels the UK to return its emissions of GHGs to their 1990 levels. On the basis of current trends this will require a reduction of some 10 million tonnes of carbon (mtC) relative to baseline emissions. This represents the contribution of the UK to slowing the pace of anthropogenic climate change.

This chapter describes the events which led up to Britain signing the Convention: noting the importance of the fact that the then Prime Minister was herself a scientist, and that the Chairman of the influential Intergovernmental Panel on Climate Change (IPCC) Working Group I was British. We draw attention to the strong endorsement given to the use of market based instruments contained in the 1990 White Paper on the environment and the reservations which were expressed at the time. We ask whether Britain was shamed into advancing existing commitments on climate change by the large number of other countries which had already announced tougher reduction targets or whether this was a response to a revision of UK carbon dioxide emissions forecasts. The prospects for future GHG emissions are considered in the light of current trends and the economic factors which are driving those trends are analysed. We calculate the relative contribution to the greenhouse effect of each of the different GHGs in terms of carbon dioxide equivalent emissions. The chapter then goes on to describe how the Government has attempted to meet its commitment to the Climate Convention and argues that the measures which were announced were not consistent with earlier indications about how the Government might choose to tackle the problem. We evaluate different policies in terms of their potential contribution to achieving the carbon savings and consideration given to the overall cost effectiveness of the policies which have been put in place. Emphasis is placed on unearthing the party political considerations which helped to shape the UK Government's policy response and on explaining the unpopularity of the decision to impose VAT on domestic fuel. Finally, we speculate on the prospects for a UK carbon tax.

UK POLICY ON CLIMATE CHANGE: THE ROAD TO RIO

The UK's interest in the problem of global climate change dates from its signature of the Montreal Protocol to phase out Chlorofluorocarbons (CFCs) and other ozone depleting substances in 1987. CFCs were at that time implicated in both ozone layer depletion and global warming.[2] The following year the DoE published its own preliminary assessment of the effects of climate change on the UK environment in June (DoE, 1988). Its ironic reference to the fact that many of the UK's holiday beaches would disappear just as the climate of southern and eastern England was becoming like that of Biarritz was widely reported in the newspapers. In the same month 300 participants attended a conference in Toronto. Whilst its declared target of a 20 per cent reduction on 1988 carbon dioxide emissions by 2005 is often referred to as the Toronto Agreement, the conference had no official status, and neither was its target for reductions much more than a figure plucked out of the atmosphere itself. The real significance of the Toronto meeting lay in that it began to look as if there was some scientific consensus on global warming.

Grasping the public mood, in September of that year the Prime Minister, Margaret Thatcher, made an important speech to the Royal Society in which she demonstrated both her understanding of the science which underlies the greenhouse effect and her commitment to sustainable development. She spoke of the UK's contribution to climate research and outlined the importance of scientific study in placing environmental policy on a firm footing: 'We have an extensive research programme at our Meteorological Office and we provide one of the World's four centres for the study of climatic change. We must ensure that what we do is founded on good science to establish cause and effect.' She went on to say that: 'the health of the economy and the health of the environment are totally dependent upon each other ... Protecting this balance of nature is therefore one of the great challenges of the late Twentieth Century.'

The IPCC was formally set up in 1988 with a British scientist, Dr. John Houghton, as the Chairman of Working Group I. That the other two Working Groups were chaired by an American and a Russian respectively illustrates the relative importance of the contribution made by British scientists. In March 1989, a conference in the Hague hosted by the Netherlands, Norway and France produced the Hague Declaration. This called for legal controls to be put on GHG emissions. But the unanimous call for a Climate Convention masked growing disagreements, with Japan, the former Soviet Union and the USA all expressing doubts about the need for immediate action. Meanwhile, in the UK, in its first ever environment

White Paper the UK Government declared that global warming was one of the biggest environmental challenges facing the world (HMSO, 1990: section 5.1). This statement was made without any detailed assessment of the facts. Instead it reflected the Prime Minister's inclinations. As a scientist herself the Prime Minister was always prepared to listen to 'good science' (a phrase she used repeatedly at the time). That it was partly British science appealed to the Prime Minister even more. The White Paper repeated the Prime Minister's stated aim of bringing UK carbon emissions down to 1990 levels by 2005, conditional on other countries taking similar action. If the available energy forecasts could be believed the target was a fairly challenging one. These forecasts suggested that existing emissions would rise from 159mtC in 1988 to 204–211mtC by 2005. The options available to meet the target reduction hinged mainly on the expansion of nuclear power. This was argued to be the cheapest option at around zero to £50 per tonne of carbon saved and capable of providing the largest contribution (at 40–70mtC in 2020) towards carbon dioxide reduction. The potential for energy conservation was thought to be around 20mtC in 2020 at an unstated (but presumably very low) cost. An increased reliance upon gas was also visualised, providing savings of 8–35mtC at anything from £40–£110/tC. Yet nuclear power's future was already very suspect with a full scale nuclear review having being announced for 1993 or 1994 and, despite the figures, there was also no enthusiasm for conservation measures brought about by information campaigns because they had failed before. Thus a switch towards gas became imperative, but might not on its own be sufficient.

Although the White Paper did not speak explicitly of carbon taxes it strongly hinted at their possible introduction:

In the long term [the measures] will inevitably have to include increases in the relative prices of energy and fuel. This could be achieved by taxation or other means, such as tradeable permits … the Government will need to take into account the argument that market-based instruments will often be more efficient and less expensive than regulations in reducing emissions. (HMSO, 1990: section 5.25).

But the document went on to argue that: 'Long term measures affecting the relative price of energy can only sensibly be taken when competitor countries are prepared to take similar action' (HMSO, 1990: section 5.26).

The 1990 White Paper proposals on tackling Climate Change were largely influenced by the Inter-departmental Group on Environmental Economics (IGEE). IGEE was composed of leading economists from different Government departments and an academic adviser, David Pearce. Whilst the group was unanimous in recognising the role that energy prices played in reducing pollution and the relative advantages of market based instruments, a carbon tax as such was not seen by some as the ideal measure. Many of the arguments were about the sectoral incidence and in particular, the role that transport would play as one of the lead sectors in terms of bearing the burden of carbon emission reduction. Transport was singled out by some departments because of fast growing carbon emissions, and perhaps because it generally appeared to be a sector out of environmental control. It was IGEE which determined the case for 2005 rather than 2000 as the target year for returning carbon emissions to their 1990 levels, fearing that a 2000 target would be too costly. This decision was taken on the basis of emissions forecasts which were subsequently revised.[3] In October 1990, Britain agreed with its EC partners that the Community would take action aiming at stabilising total carbon dioxide emissions at the 1990 level by 2000 in the community as a whole. This target assumed that other leading countries would take similar action and recognised the targets set by individual member states such as Britain.

In November 1990 IPCC Working Group I presented its report to the Second World Climate Conference in Geneva. The report concluded that anthropogenic emissions of greenhouse gases were likely to result in an increase in temperature equal to 0.3°C per decade over the next century, and that these rapid changes could be expected to disrupt ecosystems. January 1991 saw the publication of a second report prepared for the DoE into the potential effects of climate change in the UK (DoE, 1991). This report spoke of significantly warmer winters by the middle of the next century and an increased frequency of unusually hot summers. The review group who prepared the report pointed to the possibility of changed patterns of precipitation and rising sea levels affecting coastal towns and estuaries. The greatest impact of climate change would be on the fauna and flora of the country, but there were seen to be potential benefits in terms of reduced energy demands and greater opportunities for recreation.

The first year report on the White Paper published in September of 1991 (HMSO, 1991) reaffirmed the role of a general energy price rise in meeting the year 2005 deadline, and once more singled out transport as a significant and growing source of carbon emissions.

For Britain, signing the Framework Convention on Climate Change[4] at the UNCED conference in Rio in June 1992 was in some respects just a formality, although in signing the Convention the Government also committed the UK to providing financial resources to meet the costs of emissions reductions in less developed countries. The Prime Minister (supported by

Michael Heseltine) had already announced that the UK emission target, which remained conditional on other countries taking like action, would be bought forward to 2000. The bringing forward of the target from 2005 to 2000 hardly reflected any extra determination on the part of the Government; it had much more to do with the realisation that the 1989 forecasts had been a gross exaggeration. Also, by that time virtually every other OECD nation had already committed itself to stabilisation by the year 2000 and several others, notably Denmark and the Netherlands, had gone further. Britain wanted to avoid being isolated at the conference, particularly since John Major was the first Prime Minister to announce his intention to attend.

In the second year report on the White Paper published in October 1992 the phrase 'carbon tax' appeared for the first time. The report argued:

> experience has shown that, without an external stimulus, voluntary action to improve energy efficiency tends to be limited … [price measures] can be very effective relative to the alternative options – regulation or the provision of subsidies for investment in energy efficiency … A price based mechanism [i.e., the carbon tax] allows individuals and companies to respond to the need to save energy in the way best suited to them and helps to ensure that reductions in emissions come from those who can achieve them at least cost. (HMSO, 1992: 45).

This statement leaves no doubt as to the Government's position on market based instruments. Nevertheless it was recognised that: 'before introducing a tax … the Government would need to be satisfied that it would not have an unacceptable impact on any particular group of households or on the competitive position of industry (which might simply displace CO_2 generating activities to other countries)' (HMSO, 1992).

The UK Government ratified the Convention on Climate Change in December 1993 and in January 1994 produced the UK's first report to the Conference of the Parties to the Convention as required under Article 12 of the Convention. This report included inventories of the main GHG emissions and detailed measures aimed at returning the emissions of each GHG to 1990 levels: 'the UK accepts … the commitment to take measures aimed at returning emissions of Greenhouse Gases to 1990 levels by the year 2000. The UK has prepared … measures designed to achieve this commitment for *each* of the main greenhouse gases' (HMSO, 1994: section 1.2, emphasis added). In fact, the Convention did not require the emission of each GHG to be returned to 1990 level; rather, the signatories to the Convention agreed upon 'the need for developed countries to take immediate action … towards comprehensive response strategies at the global, national and, where agreed regional levels that take into account all greenhouse gases, *with due consideration of their relative contributions to the enhancement of the greenhouse effect*' (UNCED, 1992: Article 1, para. 18; emphasis added).

TRENDS IN UK GHG EMISSIONS

With all the talk of returning carbon emissions to their 1990 levels it is as well to remember that UK emissions of carbon actually fell from a high point of 182mtC in 1970 to 159mtC in 1991[5], although this is not a trend which is expected to continue. The fall in emissions appears to be driven mainly by a decline in emissions from the industrial sector, by 47 per cent over the period 1970–91. Domestic emissions and emissions from power stations have also been declining. These reductions are partially offset by a very rapid rise in emissions from road transport, which over the same period rose by nearly 88 per cent and now accounts for 19 per cent of all emissions. UK emissions represent approximately 2.2 per cent of global emissions.

Carbon emissions by type of fuel have changed radically over the last 20 years. In 1970 solid fuel accounted for 100mtC, or 55 per cent of total emissions. By 1991 the relative contribution of solid fuel had fallen to 41 per cent and in absolute terms to 66mtC. Over the same period carbon emissions from gas have risen from 10mtC to 32mtC or, in relative terms, from 5.5 per cent to 20 per cent. Newbery (1993) argues that the demise of the coal industry can in part be traced back to the Large Combustion Plant Directive signed by Britain in 1988. Although intended to reduce sulphur emissions the Directive will almost certainly limit emissions of carbon from UK power stations too by inducing fuel switching despite the fact that flue gas desulphurisation (FGD) increases carbon dioxide emissions. By 1993 over 9GW of gas fired capacity with the potential to displace 28mt of coal was either operational or under construction (the so called 'dash for gas'). Without the dash for gas UK carbon emissions would be considerably higher by 2000. But there are limits to the amount of fuel switching which can occur and several of Britain's old Magnox nuclear reactors are reaching the end of their useful working lives. Carbon emissions from the electricity supply industry might well start to pick up by the end of the decade.

The government forecasts that by the year 2000, in the absence of any policy changes UK emissions will reach 170mtC (see Table 8.1)[6] Barker *et al.* (1993) have produced independent estimates of future carbon emissions, suggesting that emissions would most likely be below 1990 levels in the year 2000 even in the absence of policy intervention.

The UK and Global Warming Policy

Table 8.1   UK carbon emissions projections (1990–2005)

	1990	1995	2000	2005
Government				
Households	41	39	41	42
Industry/Agriculture	56	56	58	61
Commercial/Public	24	23	26	30
Transport	38	41	45	49
Total	159	159	170	182
Barker *et al.*				
Total	160	152	157.9	166.7

Sources: DTI (1991) and Barker *et al.* (1993). Recently the time series data on carbon emissions have been revised downwards to take account of new measurements regarding the carbon content of North Sea gas. This has had the effect of reducing 1990 emissions estimates from the 160mtC used as the basis for the DTI's EP59 document to 158mtC. The figures in this table do not reflect that revision.

In contrast to carbon, methane emissions have declined only modestly since 1970. The single most important source of methane emissions in the UK is the deep mining of coal. A large dip in overall emissions can be observed for the years 1984 and 1985 covering the miners' strike. Emissions from this source mirror the general decline in the coal industry over the same period and have now been halved. Unfortunately, the decline in emissions from the deep mining of coal has been balanced by increases in emissions from virtually every other source. Emissions from landfill sites (which remain extremely uncertain) and leakages of gas have both increased greatly. A data series for nitrous oxides is not currently available and the emission sources themselves are still uncertain, but almost all of the 175 000 tonnes are accounted for either by industrial emissions or by emissions from the soil. Using 100-year global warming potentials (GWPs)[7] of the different gases it is possible to calculate that Britain's emissions of carbon dioxide, methane and nitrous oxides were equivalent, in terms of warming potential, to 182mtC. Carbon emissions themselves contributed 87 per cent of total emissions.

MEASURES TO CUT UK CARBON EMISSIONS

The Government has announced a wide variety of measures to cut UK carbon emissions, the combined effect of which it is hoped will reduce emissions by 10mtC relative to the year 2000 baseline. This section critically analyses the main measures and considers why a carbon tax, which many would have judged to be the preferred option, was not adopted.

VAT on Domestic Energy Use

The decision to impose VAT on domestic fuel and power at 8 per cent from April 1994 and then at the full rate (currently 17.5 per cent) from April 1995 caused an outcry because of its adverse distributional implications. Energy is a necessity in the sense that its share in total spending falls as income rises even though richer households spend absolutely more on energy (see Table 8.2). There is also evidence that pensioner households spend proportionately more than non-pensioner households on energy and that Scottish households spend proportionately more on energy than do English or Welsh households. For these reasons imposing VAT on domestic fuel and power is a highly regressive step. Giles and Ridge (1993) calculate that under the new arrangements the average increase in VAT payable by households in the lowest 10 per cent income bracket is likely to be £108 per year from April 1995, whilst for the top 10 per cent of households it is £135. Crawford *et al.* (1993) simulate the distributional impact of a 17.5 per cent increase in the price of domestic energy (see Table 8.3) and find that whilst the average fall in energy consumption is 5.8 per cent it is the poorest households which seem to economise on energy the most.

Table 8.2   UK average budget shares for domestic fuel and motor fuel expenditure by income decile group (1988)

Income group	Domestic fuel	Motor fuels
1	15.5	0.5
2	11.6	1.2
3	9.1	2.1
4	7.2	2.7
5	6.3	3.5
6	5.7	3.7
7	5.1	3.9
8	4.7	3.8
9	4.2	3.9
10	3.9	3.3
All	7.3	2.9

Source: Smith (1992), Table 2.

Table 8.3   The short run effects of imposing VAT at 17.5% on households of different income

Income quintile	Additional tax paid per week	Additional tax as % of spending	% change in energy consumption
1 (Poorest)	£1.56	2.0	−9.2
2	£1.83	1.3	−8.3
3	£2.11	0.9	−6.2
4	£2.18	0.7	−4.2
5	£2.63	0.6	−1.1
All households	£2.06	1.1	−5.8

Source: Crawford *et al.* (1993), Table 4.1.

The anger which followed the announcement was not diminished by the explanation that some such measure was necessary to meet the UK' obligations towards the Climate Convention or by assurances from the then Chancellor, Norman Lamont, that an adequate compensation package would be put in place for vulnerable groups. Such a step would obviously conflict with the government's desire to raise revenue, and imposing VAT on fuel would contribute £2.3 billion in revenue to the Treasury in the financial year 1995–96 and cut the PSBR, which was forecast to rise to £50.1 billion in the year 1993–94. Besides, the Government had already broken an election pledge not to raise VAT. Seemingly contradictory messages from different Government ministers about how much compensation might be forthcoming added to the confusion.

If the Government had hoped noiselessly to raise a large amount of tax revenue from this source under the guise of a 'green' measure then the policy backfired.[8] Although the arguments which followed the Chancellor's announcement revolved around how much compensation would be required to protect vulnerable groups, it was in fact far from obvious that any deliberate act of compensation would be needed. Currently more than 15 million people draw means tested benefits or pensions which are automatically uprated in line with the Retail Price Index (RPI). Energy prices are a constituent part of the RPI. However; benefits are uprated annually, usually some time after price rises have actually occurred. Thus there is an argument for providing temporary relief. Furthermore, the RPI is based on the spending patterns of a representative household and fails to reflect the spending patterns of certain groups

hose on low incomes and pensioners consistently spend more on domes-
ic energy than equivalent households at the same income level. These
households need extra help.

The compensation scheme for VAT on domestic fuel, outlined by the
Chancellor, Kenneth Clarke, in his November 1993 budget, involved
naking special payments to those on means tested benefits as well as
ncreasing in the state pension. Clarke proposed to calculate the increase in
ncome related benefits which would have been payable at the beginning
f April 1995 and to pay it a year early. The same procedure is to be
dopted the following April. In April 1996 the payment is to be made per-
nanent. Clarke also recognised that for pensioner households the RPI
nderstates the importance of fuel. He therefore proposed to give those
ingle pensioners on income support an extra 50 pence a week (in addition
o the uprating of their benefits) from April 1994 and for married couples
n additional 70 pence. From April 1995 there will be the same increases
gain when the VAT rate rises to the full 17.5 per cent. Chancellor Clarke
lso declared his intention to provide all other pensioners the same extra
elp with their fuel bills as that afforded to pensioners on income support.
Cold weather compensation payments were increased from £6 to £7 a
veek, starting in the 1993–4 winter and to £7.50 a week the following
vinter. Clarke also added an extra £35m to the Home Energy Efficiency
Scheme to help insulate the homes of pensioners. The cost of the compen-
ation package is expected to be £416m in the financial year 1994–95,
£897m in the year 1995–96 and £1.3bn in the year 1996–97.

The impact that the VAT change will have on carbon emissions
depends primarily upon the matrix of 'own' and 'cross' price elasticities
f the household demand for fuel.[9] The price changes may have a lagged
mpact due to the slow turnover in the stock of households' durable goods
o that the eventual changes in carbon emissions may not be fully felt even
y the year 2000. Measures taken to reduce the carbon emissions from
lectricity generation also lessen the impact of the VAT changes on
verall emissions. According to the Government, imposing VAT could
ave 1.5mtC annually by 2000 (HMSO, 1994). Barker (1993) puts the
ransient impact until 2000 somewhat higher at 2.9mtC.

Increased Road Fuel Duty

Chancellor Lamont increased the duty on road fuel by 10 per cent in his
March 1993 budget with a commitment to raise duties by an average of 3
per cent annually in real terms in future budgets. This was expected to
produce savings of 1.5mtC by the year 2000 (HMSO, 1994) and to raise

£1.02 billion additional revenue in 1995–96. In the November 199?
budget Chancellor Clarke raised the duty on a gallon of petrol again and
vowed to increase the duty on road fuel by at least 5 per cent annually in
real terms. Prior to the March 1993 Budget the average price of a gallon
of unleaded four star petrol was 216.4p. Minimum adherence to the price
increases promised should see the price of petrol rise to around £2.81 per
gallon in real terms by the year 2000: an overall price increase of around
30 per cent.

Using a model of the demand for road transport fuel Virley (1993) has
independently estimated the impact of fuel price increases on carbon
emissions. Virley's model focuses on the short run dynamics of the
adjustment to higher prices and confirms that strict adherence to the
minimum commitments made by Chancellor Lamont would reduce
carbon emissions by 1.5mtC by the year 2000 with an uncertainty range
of 0.6–2.2mtC.[10] This should be compared to a baseline in which carbon
emissions grow from 29.9mtC in 1990 to 36.5mtC by 2000. The extra
increases announced by Kenneth Clarke will reduce carbon emissions by
an additional 1mtC.

That the Government singled out transport for special attention should
come as no surprise given the strong signals contained in the environment
White Paper. In sharp contrast to VAT on domestic energy, road fuel duty
is a progressive tax (see Table 8.2). There are also economic arguments in
favour of increasing the cost of motoring. Pearce and Maddison (1993)
for example argue that due to a large discrepancy between the social costs
and benefits of road transport, taxes on road transport should rise in any
event. Whilst road fuel duty alone cannot reflect the marginal social costs
of each journey, an increase in duty may bring about desirable changes
such as a reduction in the volume of traffic, decreased accidents, noise and
air pollution. Eliminating congestion through road pricing would further
encourage people to switch the time of their journeys and decrease carbon
emissions by reducing the prevalence of stop-start driving conditions
which are so detrimental to fuel consumption (see Hughes, 1990). Apart
from reducing congestion costs, road pricing might also be effective in
reducing carbon emissions insofar as it encourages a switch to mass
transit, given the lower carbon emissions per kilometre for passengers
travelling by public transport.

On the debit side, however, the regional incidence of road fuel tax
increases suggested other concerns, with a more than proportionate burden
of the tax falling upon isolated rural communities. There were also concerns regarding the impact of increases in road fuel duty upon the price of
delivered goods.

Other Measures: Standards, Information and Exhortation

Despite the unenthusiastic billing given them by the environment White Paper, the majority of the measures announced by the Government seem to rely upon standards or direct regulations, subsidies, the provision of information and moral exhortation. Although it is difficult to tell in advance (or even in retrospect) the impact that these measures will have on carbon emissions, the Government has accredited them with savings regardless. This section comments critically on the cost effectiveness of such measures.

In order to reduce domestic energy demand the Government is pressing for minimum efficiency standards to be adopted as soon as possible at an EU level. It has already set standards for all new buildings, a step which is projected to reduce carbon emissions by 0.25mtC. The problem with a standards based approach is that consumers are heterogenous and mandatory standards therefore risk imposing very high costs on some of them. And whilst it is true that a 10 per cent increase in the average energy efficiency of domestic appliances would save £100m worth of electricity, these savings have to be compared with the increased costs of producing the domestic appliances and in some cases a decrease in their usefulness around the home.

The Energy Saving Trust was set up in 1992 by British Gas, Regional Electricity companies and the Government to provide 'well targeted incentives' (a nice euphemism for subsidies) to encourage the take-up of energy efficient devices. The Energy Saving Trust has adopted the goal of achieving reductions of 2.5mtC by 2000. Schemes announced so far include the provision of grants for the purchase of high efficiency gas condensing boilers and the installation of combined heat and power systems in residential properties. Another scheme provided financial incentives to the purchasers of low energy light bulbs. The objection to the use of such subsidies to control pollution is not that they do not have the desired effect: rather it is because the total cost of achieving pollution cutbacks in that manner is likely to be substantially higher than could be achieved by using a pollution tax. Even 'well targeted' subsidies invariably end up supporting the wrong set of technologies.

The Government is also counting on energy efficiency labelling to reduce household energy demand but, in contrast to energy price rises, there is no evidence to suggest that campaigns of this sort have ever produced significant results. And anyway, the state of energy efficiency labelling in the UK is such that only for refrigerators and freezers will a scheme be in operation by early 1994. The UK is also eco-labelling to inform the consumer of the environmental impact of a household good over its lifetime. The first product criteria for washing machines and dishwashers were agreed in June 1993. However, why informing consumers about the environmental impact of their purchases should necessarily lead to a change in their behaviour is unclear. Individual action has a negligible impact on global climate change and voluntary restraint is not compatible with private incentives. A handful of individuals may derive moral satisfaction from reducing carbon emissions, but it seems rather unwise to rely on this. The Government has also started a campaign called 'Helping the Earth Begins at Home'. This campaign seeks to provide advice on energy saving in the home whilst emphasising the role of the individual in reducing national carbon emissions. Once again there is no evidence that information on energy saving will reduce energy demand or that attempts to establish a sense of individual responsibility will succeed.

Pure exhortation also appears to form the backbone of the 'Making A Corporate Commitment' campaign. This campaign is intended to reduce carbon emissions from the corporate sector. Individual companies commit themselves to publishing a corporate policy and establishing an energy management structure whilst simultaneously raising awareness of energy efficiency among employees. Companies set themselves performance improvement targets and monitor performance levels which are then reported to shareholders, board members and employees. Sixteen hundred companies have signed up to the campaign already, but the commitments required are vague and non-binding. Of course some businesses will declare targets for reducing carbon emissions in order to create a green image with their customers but, in the vast majority of cases, these reductions would have been undertaken anyway because they were privately profitable. The absence of any pressure placed on business reflects the anxieties of the government with regard to the competitive position of domestic industries. Overall it is hard to see the campaign making much of a difference.

In April 1992 the Government launched a new Energy Management Assistance Scheme (EMAS) to provide financial help to companies with less than 500 employees, enabling them to obtain consultancy advice on the design and implementation of energy efficiency projects. The explanation given for this measure is that, unlike larger organisations, many small companies do not have a management structure which can appraise energy management issues. In 1993–4 EMAS was expected to pay grants totalling £1.7m to over 3000 small businesses in the UK. The Government anticipates that this will promote annual energy efficiency savings of five times the expenditure on grants. Many investors would be suspicious of a scheme which claimed to offer a 400 per cent rate of return! It will in any event be difficult to verify the extent to which these energy savings were additional and would not have occurred anyway. Energy efficiency advice is already available to business from energy consultants. Whilst many small businesses would not hire an energy consultant because the energy savings which they could secure would be outweighed by the cost of hiring such a professional, it is important to remember that the time of an energy consultant represents a real resource cost to the economy.

The Government has also declared in the Government White Paper on coal its acceptance of a report by the Renewable Energy Advisory Group, which stated that the contribution of renewable energy to electricity generation should rise to 1500MW by the year 2000 rather than the previously accepted objective of 1000MW. This might save 0.5mtC annually. And finally, the Government has declared a target for the public sector of reducing energy demand by 20 per cent relative to 1990 levels which might save 1mtC (see Table 8.4).

Table 8.4   Sources of saving on carbon emissions in the UK

	Carbon saved: sector government estimates (mtC)
Energy consumption in the home VAT on domestic energy use Energy Saving Trust Energy efficiency advice Eco labelling EC SAVE programme Improved building regulations	4
Energy consumption by business Making a corporate commitment Best practice programme Energy Efficiency Offices Energy Management Assistance Scheme Energy Saving Trust Energy design advice scheme EC SAVE programme Improved building regulations	2.5
Energy consumption in the public sector Targets for public sector bodies	1
Transport Road fuel duty increases	2.5
Total	10

Source: HMSO (1994).

The European Carbon Energy Tax

The EC proposed that a Europe wide tax should be introduced at a rate of $3 per barrel of oil equivalent in 1993, rising by $1 per year to $10 in the year 2000. The tax rates were to be allocated across all fuels (except for renewables) half to the carbon content of the fuel and half to the energy content. The proposals exempted certain energy intensive industries from the tax (thereby assuring their competitiveness) until some more general tax was adopted across all the OECD nations.

Many would have judged such a tax to have been the best available option. Simulations of a macroeconometric model performed by Barker *et al.* (1993) suggested that the EC tax would have reduced UK carbon emissions by 10.7mtC relative to the year 2000 baseline. The carbon energy tax proposal effectively addressed concerns regarding international competitiveness by exempting energy intensive industries, and the decision to press ahead with more traditional policies seemed to be at right angles to the stance taken in the environment White Paper. Why was the carbon energy tax proposal rejected when the Government had previously declared that: 'Britain [will limit carbon dioxide emissions and] will work positively within the EC on proposals for common measures where action is best taken on a Europe wide basis' (HMSO, 1991: section 3.35)?

Chancellor Clarke argued against the tax ostensibly on the grounds of competitiveness: that the carbon energy tax would harm the job creating sectors of the economy. But in Barker's simulations the tax had a small but positive impact upon gross domestic product (GDP) and unemployment actually fell by 50 000 relative to the baseline. The Government elsewhere referred to the macroeconomic impact of the carbon energy tax as 'slight' (HMSO, 1992: 45). Rather it appears that attempts to establish an EC wide policy on global warming coincided with a period of conflict over moves to greater political and economic union and the EC carbon tax broached the highly sensitive subject of EC influence over national taxation issues. Disagreement over the ratification of the Maastricht Treaty had almost brought down the Government just a few months earlier in November 1992. The British Government was concerned about the symbolic dimensions of any decision on an EC wide tax. The EC carbon energy tax proposal had to be sacrificed. Chancellor Lamont did not rule out future coordinated policy but said in his March 1993 budget speech that he was unwilling to set a precedent for tax measures to be determined by the European Parliament: 'I remain unpersuaded of the need for a new European Community tax. Tax policy should continue to be decided here in this house – not in Brussels.'

A secondary factor leading to the rejection of the carbon energy tax was that earlier in October 1992 the Government had announced the closure of 31 out of 50 of British Coal mines. This announcement caused uproar and prompted a humiliating reversal by the Secretary of State for Trade and Industry, Michael Heseltine. Heseltine announced a moratorium on the pit closures, and subsequently in the pits review a plan to keep some of them open. The public reaction to the pits closure programme made it more difficult to impose a carbon tax since coal has a high carbon content relative to other fuels and would have been taxed more heavily. A fear of the political consequences of further pit closures, but mainly the need to be seen to retain control over matters of domestic taxation, prevented the Government from adopting the EC's carbon-energy tax.

MEASURES TO CUT NON-CARBON GREENHOUSE GAS EMISSIONS

Up to this point the focus of attention has been on reducing carbon emissions, but the decision to adopt a rigid gas by gas approach means that the Government also has to return emissions of each non-carbon GHG to their 1990 levels. This section briefly considers the prospects for achieving such reductions for methane and nitrous oxide.

The Government believes that a combination of economic trends (the decline of the coal industry) and policies directed at landfills will reduce methane emissions by 0.6mt relative to their 1990 levels. This reduction is equivalent to 1.8mtC. Nitrous oxide emissions reductions have an especially important role to play since the GWP of one tonne of nitrous oxide is equivalent to that of 270 tonnes of carbon dioxide. The main source of nitrous oxide emissions in the UK is industry, mainly from adipic acid which is an intermediate in the production of nylon. Emissions from this source amounted to 80kt of nitrous oxide in 1990 when there were two adipic acid plants operating in the country. Du Pont (UK) Ltd, the owner of both plants, are now committed to reducing emissions of nitrous oxide by means of catalytic destruction beginning in 1996, presumably to forestall possible regulatory measures. Trials have shown that this will result in emissions falling to 0.02kt per year (depending on operating conditions). On a carbon dioxide equivalent basis this is the equal to a reduction of 6mtC, which goes a considerable way to returning carbon dioxide equivalent emissions to their 1990 levels. But after this steep decrease, nitrous oxide emissions may begin to rise following the decision which requires all new cars to be fitted with catalytic converters. Although procuring an improvement in urban air quality these devices tend to cause a 3–5 fold increase emissions of nitrous oxide from vehicles.

CONCLUSION

Forecasts of future UK carbon emissions are very uncertain and have been revised downwards on more than one occasion. The Government currently forecasts that business as usual emissions will rise from 160mtC in 1990 to 170mtC in 2000. In order to meet its commitments the Government has announced measures which it maintains will reduce carbon emissions by around 10mtC by the year 2000, as well as policies to take the emissions of other GHGs below their 1990 levels. Judging the extent to which any of the measures announced by the Government will succeed in reducing carbon emissions is extremely complex. The combined effect of the VAT and road fuel duty increases appears to be in the region of 4–4.6mtC. The impact of other schemes, such as the information campaigns and attempts at moral persuasion, defy quantification altogether. No doubt some consumers will derive moral satisfaction from reducing domestic energy consumption even if energy prices do not change, but it seems unwise to place too much faith in this. A few schemes appear to be positively naive in failing to understand the private incentives of the individuals involved.

The real reason for imposing VAT on domestic energy was the need to raise tax revenue. The VAT announcement was greeted with hostility because it was recognised for what it was, and the Chancellor was not believed when he said that vulnerable groups were to be compensated. This disbelief reflected the Government's general lack of credibility and the contradictory signals as to just how much compensation was to be forthcoming. It is perhaps unfortunate that the decision was not taken to tax road fuel even more heavily. There is strong evidence that the social costs of road transport exceed the social benefits, and that fuel duties should rise in any event. A more complete attack on the transport problems of the UK requires the use of road pricing and this might reduce emissions from road transport even further.

There was at one time a strong commitment in Government to the use of market based instruments for controlling pollution. Indeed, the likelihood of carbon taxes had to be built into the prospectus for the privatisation of the electricity industry for fear of subsequent legal action if such taxes harmed the prospects of shareholders in the new electric utilities. But the range of measures which have been announced to reduce GHG emissions do not seem to reflect this earlier resolve. An interesting question therefore, is whether the underlying commitment to market based instruments (such as the carbon tax) has changed from the clear signals given in the White Papers, or whether the Government has run up against political constraints. In this chapter we have tried to show that UK policy on global warming was shaped by political circumstances. Other developments in the sphere of acid rain policy suggest that market based approaches are still favoured in respect of tradeable permits for nitrous oxide and sulphur dioxide emissions.

The main reason the UK Government stood out against the European Commission's proposed carbon energy tax was because of the deep fault lines in the Conservative Party over the vexed question of closer European ties. The ratification of the Maastricht Treaty almost brought down the Government. The Government could not afford the appearance of letting matters of fiscal policy be decided by Brussels. The EC carbon energy tax had to be sacrificed. Concerns over the impact of a carbon energy tax upon the coal sector reinforced this view. The negative impact on the coal industry of a carbon tax would have been devastating. Fear of the consequences of closing more mines and an unwillingness to devolve any more power to Brussels hamstrung UK global warming policy.

Although a carbon tax is not currently needed, given the measures which have already been announced, two external pressures remain. First, the carbon dioxide target is a short term target only and there are limits to what can be achieved through energy efficiency campaigns and increases in road fuel duty. The UK Government has already indicated its concern about what happens after 2000 and has left the door open for future coordinated action. Second, the European carbon energy tax proposal has been re-packaged by the European Commission as an employment measure, the argument being that the fiscal neutrality provision in the proposal can be used to cut taxes on labour. Carbon taxes are down, but not out.

Notes

1. This material may not be cited, reproduced or quoted without the permission of the authors. Helpful comments made by participants at the 1993 Colloquium on UK Environmental Policy in the 1990s hosted by the University of Newcastle, and other members of CSERGE, are gratefully acknowledged. All errors remain the responsibility of the authors.
2. CFCs are now thought to be broadly neutral with respect to climate change since their indirect effect is to destroy stratospheric ozone which is, ironically, itself a potent GHG.

3.   The revised forecast was for baseline emissions of 156–178mtC by the year
       2000. This revision occurred partly because the previous forecast was
       based on economic growth assumptions which turned out to be overly
       optimistic.

4.   The Framework Climate Convention has been signed by 168 countries and,
       having achieved its 50th ratification, came into force on 21 March 1994. Its
       ultimate goal is 'the stabilisation of GHG concentrations in the atmosphere
       at a level which will prevent dangerous anthropogenic interference with the
       climate. Such a level should be achieved within a time frame sufficient to
       allow ecosystems to adapt naturally to climate change, to ensure that food
       production is not threatened and to enable economic development to
       proceed in a sustainable manner.' Whilst all signatories are to prepare
       national inventories of GHG emissions and sinks, the Convention states that
       the climate system is to be protected with regard to the differentiated capa-
       bilities of states. Accordingly, developed countries must be prepared to take
       a leading role. Developed countries are required to provide to the
       Conference of the Parties details of measures to individually or jointly
       reduce GHG emissions back to 1990 levels by the year 2000 (GHGs
       covered by the Montreal protocol, primarily CFCs, were made an excep-
       tion) within six months of the date of the convention coming into force. A
       certain degree of flexibility was afforded to the economies in transition in
       respect of meeting the target. Developed country parties (excluding the tran-
       sition economies) are further required to provide financial resources to meet
       the 'agreed full incremental costs' of implementing a range of measures in
       the developing country parties (UNCED, 1992).

5.   These figures exclude carbon emitted from air travel (which is a rapidly
       growing sector) and offshore oil and gas. Provisional estimates suggest that
       including offshore oil and gas could increase UK emissions by between 4
       and 6mtC for the year 1991, but there is uncertainty as to whether these
       emissions should be regarded as part of the UK emissions inventory or not.

6   Due to uncertainty regarding the rate of economic growth and the future
       path of fuel prices, a reduction in emissions for 2000 from 157 to 179 MtC
       is possible.

7.   GWPs measure the summed direct radiative forcing of a gas over a fixed
       lifetime relative to that of CO_2. They do not include the indirect effects of
       the gases and the time horizon over which the gas is integrated can affect
       the results substantially (see Brown and Adger, 1993).

8.   It is in fact possible to construct a case for ending zero rating of energy on
       the grounds of economic efficiency. There was a belief in some policy
       circles that imposing VAT on domestic energy was desirable on these
       grounds.

9.   The own price elasticity for fuel is the percentage change in the demand for
       that fuel given a 1 per cent increase in its price. The cross price elasticity of
       demand for fuel is the percentage change in demand given a 1 per cent
       increase in the price of a different good (e.g., another fuel).

10.   There must, however, be some doubt as to whether motorists will react to
       the announcement of future price changes in the same way that they react to
       unanticipated price changes.

References

Barker, T. (1993), 'VAT the Imperfect Way to Tackle Carbon Dioxide Emissions', *New Scientist*, 1866, 6.

Barker, T., Baylis, S. and Maben, P. (1993), 'A UK Carbon Energy Tax: The Macroeconomic Effects', *Energy Policy*, 21, 296–308.

Brown, K. and Adger, N. (1993), 'Estimating National Greenhouse Gas Emissions under the Climate Change Convention', *Global Environmental Change*, 3, 149–58.

Crawford, I., Smith, S. and Well, S. (1993), *VAT On Domestic Energy*, Institute of Fiscal Studies Commentary No. 39.

DoE (1988), *Possible Impacts of Climate Change on the Natural Environment in the United Kingdom* (London: DoE).

DoE (1991), *The Potential Effects of Climate Change in the United Kingdom* (London: HMSO).

DTI (1991), *Energy Paper 59: Energy Related Carbon Emissions in Possible Future Scenarios in the United Kingdom* (London: DTI.)

Giles, C. and Ridge, M. (1993), 'The Impact on Households of the 1993 Budget and the Council Tax', *Fiscal Studies* 14, 1–20.

HMSO (1990) *This Common Inheritance* (London: HMSO) Cmnd 1200.

HMSO (1991) *This Common Inheritance: The First Year Report* (London: HMSO) Cmnd 1655.

HMSO (1992) *This Common Inheritance: The Second Year Report* (London: HMSO) Cmnd 2068.

HMSO (1994) *Climate Change: The UK Programme* (London: HMSO) Cmnd 2427.

Hughes, P. (1990), *Transport and the Greenhouse Effect* (Milton Keynes: The Open University Press).

IPCC (1990), *Climate Change: The Scientific Assessment.* (Cambridge: Cambridge University Press).

Maddison, D. (1994), 'Is the Climate of Central England Changing?', *The Optimal Control of Global Warming*, PhD Dissertation, Strathclyde University (forthcoming).

Newbery, D. (1993), 'The Impact of EC Environmental Policy on British Coal', *Oxford Review of Economic Policy*, 9, 1–28.

Pearce, D. W. and Maddison, D. (1993), 'Transport and the Environment', in Pearce D. W., Turner R. K., O'Riordan T., Adger N., Atkinson G., Brisson I., Brown K., Dubourg R., Frankhauser S., Jordan A., Maddison D., Moran D., Powell J., *Blueprint 3* (London: Earthscan).

Smith, S. (1992), 'The Distributional Consequences of Taxes on Energy as The Carbon Content of Fuels,' *European Economy*, special edition, 1241–68.

Thatcher, M. (1988), *Royal Society News*, 4.

Tol, R. (1994), *The Climate Fund*, The Institute for Environmental Studies (Mimeo: The Free University of Amsterdam).

Tol, R. and de Vos, A. (1994), 'Greenhouse Statistics: A Different Look at Climate Research', *Theoretical and Applied Meteorology* (forthcoming).

UNCED (1992), *Earth Summit '92* (London: The Regency Press Corporation).

Virley, S. (1993), 'The Effect of Fuel Price Increases on Road Transport CO_2 Emissions', *Transport Policy*, 1, 1, 43–8.

9 Energy Conservation Policy

Gerald Manners

INTRODUCTION

The motives and the vigour of energy conservation policies in the UK have shifted regularly over the last 20 years. In the wake of the 1973/74 oil shock and the consequently high and volatile price of energy, the first policies to promote conservation were forged largely in response to concerns about the adequacy of future energy resources. Exhortations to 'save' energy were paralleled by regulations that sought to prevent its unnecessary waste in public buildings and a modest subsidy for loft insulation in the home. Overseen by a small division within the Department of Energy, the programmes had modest resources and a limited visibility that reflected their low political priority. In the early 1980s government interest in the issues initially waned until the considerable economic advantages of energy conservation investments came to be more widely appreciated, and the notion of 'energy efficiency' was born. The Energy Efficiency Office (EEO) was established within the Department of Energy in 1983 to support a more vigorous marketing of conservation investments, and – especially with 1986 declared to be Energy Efficiency Year – policy was briefly given enhanced resources and a high public visibility led by the then Secretary of State for Energy, Peter Walker.

As environmental concerns shifted towards the end of the 1980s from the harmful consequences of energy production to the adverse effects of energy consumption, the potential contribution of energy conservation policies to the amelioration of fears about acid rain and global warming was initially ignored. It is arguable that the reduced resources and lower visibility of the EEO at this time was a consequence of the Government's preoccupation with the privatisation of its major energy supply interests. Only in the 1990s, in the course of the Government's preparations for and its commitments after the 1992 Earth Summit in Rio de Janeiro, were the deleterious consequences of rising energy consumption and the mounting sulphur dioxide, nitrous oxide and carbon dioxide emissions seen as being capable of substantial amelioration through improvements in the efficiency of energy use. The response of the UK Government to its obligations under the UN Climate Change Convention, together with its search for a strategy for Sustainable Development, have given a renewed political urgency to the task of reducing unnecessary energy use. In 1992, the EEO was transferred to the DoE, and today a more vigorous search is under way to determine how the environmentally harmful consequences of energy conversion and use can be most cost effectively limited through improved energy efficiency. At the same time, the broader concerns of the sustainability debate have raised new questions: in particular they have raised the issues of whether the legitimate energy requirements of the present can be met without prejudicing the satisfaction of energy needs in the future, and whether the wealth created from the exploitation of non-renewable energy resources can be used, at least in part, to ensure a legacy for future generations no less valuable than the resources withdrawn from reserves.

THE FOUNDATIONS OF POLICY

The objective of energy conservation policies is to reduce the quantity of energy consumed without in any way diminishing the quality of the services that the use of energy provides. Involving a wide range of management measures and investments in all sectors of energy demand, obvious examples are the improvement of insulation in buildings, the replacement of fixed speed by variable speed motors in manufacturing, the promotion of more energy efficient motor vehicles, white goods and other consumer durables, and the replacement of traditional tungsten lighting by low energy alternatives. All require investment in both energy management skills and appropriate hardware, but it is investment that is often handsomely rewarded. It is widely accepted that, once installed, many energy efficiency measures have a pay back period of less than one year; others have a two, three or four year pay back period. In other words, investments in thermal insulation, energy control equipment, energy system design and low energy lighting can have theoretical real rates of return of 25 per cent, 50 per cent and even 100 per cent. Even if the transaction costs of managing the process of deciding upon and implementing the investments is underestimated or neglected in such figures, such benefits contrast starkly with the 10 per cent–15 per cent real rates of return that privately owned energy supply industries normally expect, and particularly with the 8 per cent rate that the Treasury expects Nuclear Electric to apply in its investment appraisals.

The fundamental challenge of energy efficiency policies, therefore, is to redress the imbalance of investment criteria and practice that habitually characterises the supply and demand sides of the energy market. The existence of this imbalance can be explained by a multiplicity of barriers that prevent a more rational allocation of capital resources. The ignorance of consumers and the inadequacy of relevant information, the scarcity of capital in particular sectors of the market and faulty price signals, and impediments of an institutional nature (such as the landlord-tenant relationship) are now well documented, both nationally and internationally (Environment Committee, 1993; World Energy Council, 1993: 127); and they can be circumvented, if not entirely eliminated, through policy intervention. Success in so doing provides self-evident economic benefits, for individuals, business and the economy as a whole. Capital is used more efficiently; individuals and businesses satisfy their energy requirements at lower costs; the economy becomes more competitive; more jobs are created, since the installation and management of energy efficiency investments are normally more labour intensive than investments in further energy supplies; and in the medium term net energy imports will be lower than they would otherwise have been. Social and environmental benefits are also gained from energy efficiency investments. The social gains take the form of warmer and drier homes, particularly those of the lower income groups, and the consequential improvement in their occupants' health. The environmental benefits are reduced atmospheric pollution by energy users, and fewer land use problems associated with energy production.

Given the wide ranging economic, social and environmental benefits of energy conservation policies, it is not surprising to find a growing intellectual commitment by even a non-interventionist Government to the idea of accelerating a process which lies very clearly with the grain of events. After all, in 1950 the creation of each £10 000 of GDP consumed some 10 tonnes of oil equivalent energy, but by 1992 only 6 tonnes were required. Confidence in energy efficiency policies is further raised by the sheer scale of the opportunities that are available to lower energy consumption. For more than a decade now Government has stressed how cost effective energy efficiency investments could reduce the nation's energy use (and consequentially the size of its energy bills and the scale of its atmospheric pollution) by at least one-fifth (Hansard, 1983). Whilst this is arguably an understatement of the true potential of energy efficiency measures, it is nevertheless an impressive claim.

One-fifth of the 207 million tonnes of oil equivalent energy used in the UK in 1992 is equivalent to nearly 80 per cent of all the gas, or 70 per cent of all the coal, that was used in that year. One-fifth of the £50 billion spent on energy consumption in the UK in 1992 is a potential saving of £10 billion each year. And one-fifth of the 160 million tonnes of carbon that the UK emitted into the atmosphere in that year is 32 million tonnes, which is more than three times the Government's annual target for policy-induced 'carbon savings' by the year 2000. In its submission to the IPCC in 1989, the Government in fact suggested that savings of 30 per cent were potentially achievable: 10 per cent through good housekeeping and better energy management; a further 10 per cent through retrofit investments with a pay back period of up to two years: and yet another 10 per cent through investments which had a pay back period of between two and five years (DoE, 1992). Embraced as policy objectives, such savings fully qualify for the 'minimum regret' and 'precautionary principle' accolades of environmental policy debates. In that same report to the IPCC, incidentally, Government suggested that a further 30 per cent of savings was technically possible on a variety of assumptions, some of which were not wholly unrealistic taking a 20-year view.

ENERGY CONSERVATION POLICY TODAY

To achieve its environmental goals the Government is today in the process of redefining its role with regard to the promotion of greater energy efficiency in the UK. The most focused of its recent efforts was its 1992–93 public consultation exercise with regard to the reduction of carbon dioxide emissions. The significant contribution that inter-fuel substitution is likely to make to this end was subsumed in the 1992 energy consumption forecasts of the DTI that were used to estimate emission levels without further policy intervention. In consequence, the DoE's initial discussion document on *Climate Change – Our National Programme for CO2 Emissions* (DoE, 1992) was largely devoted to an exploration of the contribution that improvements in the efficiency of energy use might make to the declared objective of stabilising emissions at 1990 levels by the year 2000. Building on some of the ideas and commitments of the 1990 White Paper, *This Common Inheritance* (Secretary of State for the Environment *et al.*, 1990), particularly its Annex C, the discussion document had the virtue of elaborating for the first time by Government itself many of the measures that would be necessary to improve the country's energy efficiency. Most had been regularly urged upon Parliament and Government by the former Energy Select Committee (1991), amongst others. They were also part of a parallel agenda for action proposed by the Advisory Committee on Business and the Environment (1993)

The consultation exercise began with the DoE stressing what it saw as the central importance of 'voluntary action' – by individuals, businesses and public bodies – to improve the nation's energy efficiency. It took the stance of a Government that would prefer to remain a passive observer rather than an active participant in the management of the country's energy demands. In his Foreword, the then Secretary of State appeared to dismiss any notion that Government should espouse a central role in the promotion of greater energy efficiency: 'The more each sector of our economy can achieve through voluntary action to meet self-imposed targets, the less will be the need for Government intervention' (DoE, 1992: 1).

During the consultation process itself, however, the attention of the Government was drawn to the embarrassing legacy of only limited progress (at least by some international standards) towards the improvement of the UK's energy efficiency, a legacy that had been bequeathed by 'voluntary action' in the past. Certainly, the Government's ambition for a 20 per cent improvement in the country's energy efficiency in the decade to 1993 had not been matched by events, despite its subsequent reinforcement in a European Commission Directive. In response to these and other arguments, Government markedly shifted its ground. At a national conference on 7 May 1993, the then Secretary of State for the Environment explicitly accepted that Government must play a leading role in the promotion of greater energy efficiency; in particular, he accepted that the task of Government was to ensure an energy market environment in which cost effective energy efficiency actions and investments would be encouraged: 'Only Government can provide the basic legislative and fiscal framework for our programme, the conditions within which others will act ... It is the Government's role to provide the framework' (Secretary of State for the Environment, 1993).

Whilst the Government's acceptance of a central role in the promotion of accelerated investment in energy efficiency goods and services is a necessary condition of progress, by itself it is insufficient. Of the several other policy matters that still needed early resolution, arguably the most important concerned energy prices. Experience over the last 20 years has confirmed a clear, albeit variable, relationship between relatively high and uncertain energy prices on the one hand and a quickened pace of energy efficiency investments on the other (Energy Select Committee, 1991, para 52 ff.). At the same time, it has become ever more widely accepted – at least in theory – that mounting environmental concerns imply that sooner or later energy prices will need to reflect in some way or other the costs that their production and use impose upon the environment. Government in consequence has the task of communicating both to business and to the public generally the prospect that a rise in energy consumption taxes is inevitable if the country is to implement the 'polluter pays' principle, fulfil its international environmental obligations, and move towards a more sustainable approach to the satisfaction of energy needs. Since the mid-1980s, however, oil prices (and, as a consequence, many other energy prices) have been generally low and tending to fall in real US dollar terms. Low energy prices inevitably limit the returns to energy efficiency investments, and diminish the vigour with which possible investments are investigated. Moreover, the privatisation of the UK energy utilities has brought to many consumers not only reduced gas and electricity prices in real terms but also the expectation of further price falls in the future. Such price reductions are seen in some political circles as the litmus test of a successful privatisation programme.

Although philosophically committed to the notion of using market instruments to ameliorate environmental concerns, throughout the late 1980s and beyond the Government was reluctant to use in any significant way such financial inducements as taxes, tax concessions or subsidies to manage energy prices and thereby to accelerate the rate of investment in energy efficiency goods and services. In part this was for reasons of broader macroeconomic management, in particular the containment of public expenditure. The Government's adamant opposition to the EC's proposals for a carbon/energy tax was born of a more complex set of political considerations (especially its insistence upon the importance of subsidiarity), as well as economic arguments. The financial support provided by the DoE for improvements to the energy efficiency of low income homes – the Home Energy Efficiency Scheme (HEES) scheme – and start-up funding for consultancy advice in the small business sector – the EMAS scheme – were the exception rather than the rule. With its Budget proposals in the spring of 1993, however, the Government for the first time appeared to match its intellectual convictions with courage. The decisions to phase in the full 17.5 per cent VAT on domestic fuel and power, and to increase road fuel duties by 10 per cent and to commit Government initially to a 3 per cent increase (and subsequently a 5 per cent increase) in those duties in real terms over a number of years, represented a bold step forward. The adverse political reaction that followed the VAT decision in particular was a measure of the challenges posed by such policies: political challenges that are compounded by the relatively low elasticities of the energy demands of smaller consumers in particular, and the doubts that surround the scale of carbon savings that the DoE have suggested will flow from these taxes once they are imposed in 1994.

It was in this context that the almost simultaneous introduction of what are in effect levies on the tariff and franchise customers of the gas and electricity utilities, in order to finance the activities of the Energy Saving Trust (EST), was particularly interesting. Not only are the funds initially borrowed from the generous cash flows of the energy utilities and outside the normal tax gathering system, but their burden is hidden in minor adjustments to the tariffs charged by the utilities in their tariff and franchise markets respectively. As hypothecated taxes to pump-prime energy efficiency investments in the domestic and small business energy markets, the levies have considerable political merit. Established in 1982, the EST had its genesis in the conflict between the short term commercial interest of the energy utilities to sell as much gas or electricity as possible (provided the prices charged exceed the marginal costs of supply), and their obligation to promote greater efficiency in the use of energy by their customers. This latter obligation derives from the privatisation legislation itself, from the licenses that the utilities hold to supply gas and electricity, and from the duties imposed by the President of the Board of Trade upon their respective regulators. The potential scale of that conflict was reduced in the case of the gas industry in 1992 with the agreed revision of the tariff price formula. For the first time this included an 'E' factor which, subject to the approval of the Office of Gas Services (OFGAS), permits British Gas to pass into its tariff prices any reasonable expenditure that it might make to improve the efficiency with which gas is used by its customers: expenditure, that is, over and above that which it might incur to increase the efficiency of gas use for its own commercial reasons (such as the encouragement of fuel switching). The 'E' factor does not permit the inclusion of any measure of the profits forgone by the utility through lower gas sales as a result of the promotion of its consumers' end-use efficiency; the conflict facing British Gas is in consequence merely reduced, not eliminated.

The electricity supply industry's equivalent to the 'E' factor was proposed in July 1993 as part of the revised supply price formula. The regional electricity companies (RECs) can now be reimbursed for expenditure of up to £1 per tariff market customer for the promotion of energy efficiency. The initial reluctance of the electricity regulator to introduce such a levy – in particular, his concerns about its implicit cross-subsidy of some consumers by others – and his much narrower view of the EST initiative, was initially shared by the second gas regulator, Clare Spottiswoode. Her immediate concerns centred on the scheme's uncertain legality, and a reluctance to become increasingly involved in the administration of a hypothecated tax (*Financial Times*, 16 February 1994). This is an issue to which we return later.

In the meantime, the first (pilot) schemes to be promoted by the Trust have been launched: there is a £200 subsidy for each gas condensing boiler installed by any tariff customer; financial support for small scale combined heat and power schemes; a temporary subsidy of up to £5 on each low energy light bulb; and the establishment, on a pilot scheme basis, of a small number of local energy advice centres. The DoE has suggested that the income of the EST could rise to £300 million a year by the late 1990s (DoE, 1992: 30). Such a levy might add between 4 and 5 per cent to domestic energy bills. Lord Moore, the Chairman of the Trust, however, has proposed that a figure of £400 million a year (or a total of £1.5 billion by the year 2000) would be more appropriate, given the carbon reduction targets that have been set for the Trust by Government (Moore, 1993). Such a sum, deployed imaginatively, might be used to induce on average energy efficiency investments of at least twice that figure: whilst higher leverage rates are likely in the case of some schemes, lower expectations must be associated with others, particularly any information or education programmes the Trust might endorse, or any assistance that might be given to accelerate the installation of energy efficiency measures in low income homes. Energy efficiency investments approaching £1 billion per year in the gas and electricity tariff and franchise markets would very quickly transform the rate of progress in reducing carbon emissions from a sector of the energy use that might otherwise continue to be relatively unresponsive to environmental concerns. Given the existing and prospective scale of the Government's PSBR, in the short to medium term the EST is the only obvious means whereby a considerable uplift might be achieved in the scale of UK investment in improved energy efficiency using market instruments.

Significant improvements in the pace of energy efficiency investment will, however, require a multiplicity of measures, regulatory as well as market instruments. With regard to regulations, however, the Government has tended to move slowly and not very far. The DoE has found it acceptable to raise once again the standards of insulation required by building regulations for new properties, and through the EU's SAVE programme the minimum allowable efficiency of smaller boilers has been raised. However, the Government continues to turn its back on proposals – such as a mandatory home energy labelling scheme – that would ensure a gradual upgrading of the energy characteristics of existing buildings, which after all comprise the greater part of the country's stock. Whilst the energy labelling of appliances has been accepted as a legitimate goal, the Government has been understandably reluctant to move forward in advance of EU regulations. And although the DTI is committed to the expansion of Combined Heat and Power (CHP) generating capacity from its existing 2.3 GW to 5.0 GW by 2000, to be successful such a programme will require the removal of certain regulatory barriers to the electricity market that currently prejudice the economics of many medium and larger scale CHP schemes (Environment Committee, 1993, para 57 ff).

The Government *has* taken the point, however, that it should put its own house in order. To that end, it has committed itself to a 15 per cent reduction in its energy bills over the five years to 1996, using a '£/m^2' as well as two carbon-reduction indices (Secretary of State for the Environment *et al.*, 1994: 29). In parallel, and against the background of earlier criticisms by the Audit Commission, it has urged local authorities and such major public sector energy users as the National Health Service (NHS) to set themselves comparable targets.

The overall policy stance of Government towards the promotion of greater energy efficiency, then, is more coherent in early 1994 than it has been for some time. Driven primarily by environmental commitments – and with its goals expressed as often in terms of carbon savings as in financial rewards (see Table 9.1) – it recognises the existence of barriers to a more rational pattern of investment within the energy sector, and it has accepted a responsibility to overcome them. Nevertheless a gap persists between the theoretical acceptance by the Government of its responsibilities and the realities of their practical implementation. This gap deserves some attention.

Table 9.1   Anticipated contribution of initiatives to a reduction in carbon dioxide emissions in 2000

Initiative	*Mtc (Million tonnes of carbon)*
VAT on domestic fuel and power	1.5
Work of the Energy Saving Trust	2.5
Development of energy efficiency advice and information	1.5
Increased CHP target	1.0
Increased renewable energy	0.5
Regulatory changes (with regard to buildings and appliances)	0.75
Improvements in public sector energy efficiency	1.0
Road fuel duties	2.5
Total	11.25
Less double counting	−1.25
Net total	10.0

Source: EEO (1994).

POLICY ACHIEVEMENTS

The scale of improvements in the efficiency of energy use, and particularly those that might have flowed from policy initiatives, elude easy measurement. Whilst some progress has been achieved through changes in user behaviour – by switching off unnecessary lights, or maintaining boilers with greater care, for example – advances have been most commonly achieved through investments of many kinds. It is capital expenditure on energy surveys and associated energy advice, on improved energy control systems, on building and plant insulation, on low energy lighting, and on improved product design that provide the primary driving force towards improved energy efficiency; yet no statistical time series on investment in energy efficiency is available. One surrogate measure is the energy ratio – the relationship between energy use and GDP – which fell steadily from 10 tonnes of oil equivalent energy per £10 000 of GDP in 1950 to 6.0 tonnes in 1992 (Table 9.2). However, it is a measure that can be affected as much by structural changes in an economy (such as the contraction of older, smoke stack, energy intensive industries) as by changes in the efficiency of energy use. It can also be adversely affected by recession when lower output is not always matched by a similar reduction in energy use.

Unexpectedly, and for reasons that are only partly explicable, the ratio increased between 1989 and 1992. Expressed as an index (1985 = 100), the ratio was 93.6 in 1989 and then moved steadily up to 97.1 in 1992, a 3.7 per cent increase. It is known that investment in improved energy efficiency has been depressed over the same three years. The manufacturers and wholesalers of insulation materials and energy control equipment, for example, reported a decline in business activity of some 28 per cent

Table 9.2   Energy ratio, 1970–1992

	Ratio *(toe/£1000 GDP)*	*Index* *1985 = 100*
1970	0.85	138.3
1985	0.62	100.0
1988	0.58	94.5
1990	0.59	95.0
1992	0.60	97.1

Source: DTI (1993: 143).

between 1989 and 1992, and they saw no significant upturn in their orders in the subsequent 12 months (Association for the Conservation of Energy, 1993: 3). Although this is not surprising, given the severity of the recession generally and its distinctive impact upon the building sector in particular, it would nevertheless have been reasonable to expect a small improvement in the efficiency of the UK's energy use over this same period. Efficiency gains would have been secured from the occupancy of newly completed commercial and domestic buildings, with their improved thermal specifications; from the routine purchase of replacement capital goods and consumer products, many of which have improved energy efficiency characteristics; from the diffusion of 'state of the art' technologies, a process which proceeds even during a deep and prolonged recession; from the modest expansion of combined heat and power schemes; and as a result of the Treasury's more accommodating attitude towards contract energy management in the public sector (especially in certain key central government departments and the NHS). Whilst a lack of significant progress is easy to understand, a deteriorating energy ratio remains a puzzle.

Taking a somewhat longer view of progress, the country's record can be described as patchy. There are areas in which considerable improvements in the efficiency of energy use have been recorded: the energy-intensive industries (such as steel and chemicals) and the larger energy users in the expanding and profitable sectors of the economy (such as the supermarket chains) have responded – and continue to respond – vigorously to the economic fundamentals, and energy efficiency investments have been made to profitable effect (Secretary of State for the Environment *et al.*, 1994: 24 ff). On the other hand, medium and small scale businesses have made less impressive progress: where companies have 'made a corporate commitment' to save energy, measures and investments to reduce energy bills have often been forthcoming; but to date more opportunities have been missed in this sector than have been grasped. In the domestic market, revised building regulations have made a positive impact on the energy efficiency of new properties, but most housing is old stock and retrofit investment has been slow in both the private and public sectors. And in the public sector generally, despite many examples of excellent practice, a similarly mixed but generally disappointing picture cannot be escaped, as the recent Environment Committee (1993: para 143) and earlier Audit Commission reports have underlined. The result is that, by the more testing of international standards, the country's overall performance remains below the best (Energy Select Committee, 1991: para 22 ff). The energy ratio of Japan, for example, measured in energy use per $1000 of

GDP, is about two-thirds that of the UK. The apparent failure to turn a consensus about aims into practical achievements must be a matter for concern.

POLICY CONCERNS

Whilst the environmental commitments underlying today's energy conservation policies may have advantageously raised their public profile, they embody a trap. This follows from the Government's decision increasingly to translate environmental policy objectives into measurable physical targets. For example, the national programme to reduce carbon dioxide emissions is being developed in terms of the contribution that different policy initiatives can make towards a target reduction of 10 mt per year of carbon by the year 2000: 2.5 mt per year through the activities of the EST, 1 mt per year from additional CHP schemes, and 2.5 mt per year from rising road fuel duties, etc. (see Table 9.1). Such targets have their place in environmental politics. However, at root they are arbitrary, and in no way can they be said to reflect what is economically optimum. Indeed, they may seriously understate the scale of cost-effective energy efficiency opportunities that could be captured within the next five or six years.

Moreover, it is increasingly recognised that inter-fuel substitution is already making a significant contribution to the reduction of atmospheric emissions. Substantial reductions in carbon dioxide emissions will increasingly flow from a continuing improvement in the operating performance of the existing nuclear power stations and the (1994) commissioning of Sizewell 'B'. Further reductions will follow from the commissioning of some 12 GW of new combined cycle gas turbine power stations, which are set to capture an increasing share of base load generation at the expense of older coal-fired plant. It is not entirely clear that the scale and pace of these changes could be fully reflected in the central growth/low price projections of the now-dated Energy Paper 59 (DTI, 1992), the projections which underlie the target 'policy' reduction of 10 million tonnes of carbon emissions per year by 2000. When the effects of inter-fuel substitution are fully recognised by Government, the pressures to improve the effectiveness of its energy efficiency programmes and to ensure their full implementation could well diminish.

A second policy concern is in fact the frailty of some of the Government's policy measures. This can be illustrated by the circumstances of the EST, whose activities are central to the programme of planned reductions in carbon dioxide emissions; indeed the Trust has been asked by the Government to help reduce carbon emissions by 2.5 m per year by 2000, one-quarter of the Government's programme. The initial reluctance of the Director-General of Electricity Supply to make the promotion and assistance of consumers' energy efficiency an allowable expense within the supply tariff formula has already been noted, as has the more recent reservations raised by the second Director-General of OFGAS about the 'E' factor levy. Legislation is clearly required to clarify the regulators' position. Equally worrying is the fact that the share of the gas and electricity demands that will be subject to an EST levy is destined to contract. The threshold for gas purchases in the contract market was lowered from 25 000 therms to 2500 therms in 1992; the small business market for gas will be deregulated in 1996; and the domestic market will be fully competitive by 1998. Similarly, the franchise market for electricity consumers was reduced from those with a peak demand of 1 MW to 100 000 kw in 1994, and will be completely eliminated by 1998 (at the latest). As the number of customers in the regulated as opposed to the competitive markets for gas and electricity shrinks and then disappears, so in parallel will the number of consumers on whom a levy can be imposed and who might benefit from EST support. The Trust will almost certainly have access to some £35 million in 1994 (Environment Committee, 1993: para 38). However, its Chairman (Lord Moore) has suggested that the Department's proposal will require an annual expenditure of more than ten times that sum by the end of the decade. Given its shrinking financial base, it is unclear how the Trust will be funded in the late 1990s on a scale commensurate with the task that it has been set. One proposal is to attach a levy to the pipeline and transmission activities of the gas and electricity industries in proportion to the share of energy destined for the domestic, or domestic and small business, markets. Another is simply to add a small surcharge to customers' bills (Environment Committee, 1993: para 163); in the wake of the adverse reaction to the imposition of VAT on those bills, such a solution appears politically unlikely.

The utilities have long been seen as key vehicles for the promotion of energy efficiency ideas and action, particularly in that part of the market in which the smaller consumer is served. Partly in response to inter-fuel competition, but also as a result of political pressures, they already operate a variety of schemes to promote energy efficiency action and investment by their customers. One of the reasons why they have not initiated more in the regulated market is the lack of financial incentives for them to do so. This in turn stems from the price control formulae adopted upon privatisation, formulae which included a significant sales volume incentive. Although the 1994 revision to the supply price control for the regional electricity companies in England and Wales will reduce their dependence upon the volume of their sales to achieve reasonable profits, decisions about the more important distribution price control remain to be made. A similar change is arguably desirable in the case of the tariff market for gas market, particularly if the effectiveness of the 'E' factor arrangements and the EST are not improved (Environment Committee, 1993: para 163).

The uncertain finances of the EST is a reflection of a wider financial concern. Despite their new centrality in the environmental policy of the Government, energy efficiency programmes in general can at best be described as underresourced. Whilst the budget of the EEO has commendably been increased in recent years (Secretary of State for the Environment *et al.*, 1994), there nevertheless remains a considerable imbalance between the time, money and manpower that the Government has to date made available (and in the past has tended to spend) on supply side issues of the energy market, and the resources it remains willing to commit to the demand side of that market. The expenditure incurred by the Government in response to the coal crisis of 1992–93, and that which will undoubtedly underwrite the 1994 nuclear review, are of a different order of magnitude from those that tend to be available to the EEO.

CONCLUSION

For the full economic and environmental potential of energy efficiency improvements to be realised, Government will have to commit itself more fully than hitherto to the translation of its intellectual commitment to the promotion of energy efficiency into a more effective policy reality. In particular, it will need to place relatively less emphasis upon the straightforward task of providing more information to make the market for energy supplies and energy efficiency goods and services increasingly transparent, useful though that task may be. Increasingly, Government must give more urgent attention to the formulation and public presentation of market instruments that will encourage a quickening pace of energy efficiency investments, and qualify its instinctive opposition to further regulation in key areas of the energy market. The mandatory energy labelling of domestic properties at the point of sale is a case in point. At the end of the day, Government will have to accept the possibility, if not the likelihood, of its actions causing some disappointment for – and possibly offence to – at least some of the country's major energy supply interests.

References

Advisory Committee on Business and the Environment (1993), *Third Progress Report to and Response from the Secretary of State for the Environment and the President of the Board of Trade* (London: DoE).

Association for the Conservation of Energy (1993), *Climate Change, Our National Programme for CO_2 Emissions: A Response to the Government's Discussion Document* (London: Association for the Conservation of Energy).

DoE (1992), *Climate Change – Our National Programme for CO_2 Emissions: A Discussion Document* (London: DoE).

DTI (1992), *Energy Related Carbon Emissions in Possible Future Scenarios for the United Kingdom*, Energy Paper 59 (London: HMSO).

DTI (1993), *Digest of United Kingdom Energy Statistics 1993* (London: HMSO).

Energy Select Committee (1991), *Energy Efficiency*, Third Report, Session 1990–91 (London: HMSO).

Environment Committee (1993), *Energy Efficiency in Buildings, Vol. 1 Report*, Fourth Report, Session 1992–93 (London: HMSO).

EEO (1994), 'Measures Contributing to the UK's Climate Change Programme', unpublished paper.

Financial Times (1994), 'Off on the Wrong Foot', 16 February.

Hansard (1983), *House of Commons Debates*, 19 December, col. 42.

Moore, Lord (1993), Address to the National CO_2 Conference, London, 7 May.

Secretary of State for the Environment *et al.* (1990), *This Common Inheritance, Britain's Environmental Strategy* (London: HMSO) Cmnd 1200.

Secretary of State for the Environment (1993), Address to the National CO_2 Conference, London, 7 May.

Secretary of State for the Environment *et al.* (1994), *Climate Change: The UK Programme* (London: HMSO) Cmnd 2427.

World Energy Council (1993), *Energy for Tomorrow's World* (London: Kogan Page).

10 Constructing Regulations and Regulating Construction: The Practicalities of Environmental Policy

Elizabeth Shove

INTRODUCTION

'The only way to get us to change is to change the regulations.' With these words, house builders, architects and developers confirm conventional wisdom and, with half a smile, reaffirm their own conservatism. For them, legislation represents the only sure method of reducing energy consumption and associated carbon dioxide emissions in the built environment. Whatever their other merits, gentler strategies of persuasion and inducement have uncertain and often unpredictable consequences. And although they are usually more reliable than mere persuasion, enticing financial incentives may also fail to function as anticipated, their effects mediated by a surrounding network of priorities and perceptions of relative value. By comparison, few doubt the definitive, unambiguous power of regulatory control. Those who call for tougher legislation and tighter control do so in the belief that governments can establish, enforce, and in that way guarantee higher environmental standards if they so wish. By implication, battles surrounding the formulation of environmental policy take place within the formal political arena: once decisions have been taken and legislation enacted, industries, companies and citizens have no option but to comply.

This 'sledge-hammer' model of environmental regulation makes two important assumptions: first, that there is a clear distinction between the regulators and those they regulate and, second, that regulation is effective. These assumptions are also shared by many of those studying the evolution of environmental policy. In mapping legislative developments over time, and in comparing the codes and standards of different nations, researchers tacitly accept both the efficacy of the regulations they examine and their status as telling indicators of environmental policy.

By taking regulatory activity at face value in this way activists and observers are in danger of oversimplifying relationships between those who devise and those who are subject to environmental controls, and of overlooking the inter-dependence of policy and its implementation. Enforcement and implementation are not entirely transparent processes trailing meekly behind the real political action; far from it. Implementation is big business and as such it is affected by any number of social, organisational, economic and (little) political interests. Furthermore, the practical realities of managing the regulatory process have important consequences a long way back up the line, shaping the possibilities of control and the terms in which regulation is itself considered and conceptualised. By looking more closely at the micropolitics of environmental change we can begin to see the negotiated realities of regulation. Such an exercise suggests that environmental regulation takes place within a tangled network of inter-dependent relationships between government, industry and associated pressure groups. These relationships, which precede and inform regulatory intervention also determine the practicalities of implementation. If we want to understand environmental policy in the early 1990s then we also need to understand the regulatory contexts out of which such policies emerge and within which they are implemented.

Although focused upon the built environment, this discussion of the theory and practice of regulatory control raises issues relevant to other areas of environmental policy in the late 1980s and early 1990s. It is a good case to take. Around half of all carbon dioxide emissions relate to energy consumption in buildings, and efforts to improve energy efficiency in this sector are of significance in their own right. In addition, it is a case which clearly illustrates the delicacy and complexity of regulatory relationships.

REGULATION THEORY

Taxes, grants, guidance and regulation; these devices are all designed to change current practice and bring it into line with current policy. The regulatory option is generally seen as a last resort, used when persuasion is not enough and when other methods fail. Friendly advice and information is thus followed by clear guidance and finally by formal control. In this respect, regulation stands at the end of a linear sequence of more and less forceful intervention.

Providing that regulations are appropriately enforced, compliance is guaranteed. For this reason environmental regulations are taken to represent standards of minimum performance, standards which are in turn interpreted as a measure of the enthusiasm and commitment of the policy making community. In the building sector, as elsewhere, standards embodied in current legislation consequently define the nation's position with respect to energy efficiency (or to health and safety, pollution control, air quality, waste management, etc.). Accepting this view of regulation, careful reading of mandatory standards then shows how well, or how badly, this country compares with others.

Looking back, such analysis also reveals the rate of change in environment-related legislation. Each new edition of the thermal regulations is thus seen as an advance on the one before it. And with each new edition Britain appears to inch toward the position occupied by the Scandinavians some 30 years ago. In purely technical terms, the way ahead is seemingly well defined. Calls for government action underline this view, routinely illustrating both an unquestioning faith in the government's capacity to intervene effectively and an equally impressive certainty about the direction such action should take. This statement by Peter Smith, organiser of a 1990 conference on 'The Architect, Energy and Global Responsibility', illustrates the pattern: 'There are specific areas where ... the government must impose its Green will. By 1995 we should have raised insulation levels in building to current Danish levels. The 1990 amendments bring British regulations up to the level of Sweden in 1935' (Polan 1990). From this perspective, the country seems to be sliding along a single predetermined track of environmental/technological progress with government policy influencing the rate of movement along that line.

This linear process appears to be paralleled by another seemingly inevitable sequence of scientific certainty, according to which yesterday's risky innovations become today's uncontested standards. Figure 10.1 shows how these processes are thought to operate, illustrating both the 'natural' diffusion of innovation and associated levels of certainty and corresponding opportunities for government action.

As described here, the shape and texture of the regulatory system reflects the shape and texture of current technical knowledge and the level of policy commitment. In this account, the content and form of regulation depends upon a measure of scientific consensus and upon the level of technical knowledge within the building industry. In other words scientific uncertainty and technical competence set their own 'natural' limits, roughly defining the territory over which regulatory control might be applied. The exact scope and extent of regulation is then a question for the policy makers, for it is they who determine precisely where the regulatory lines are drawn.

What are the grounds for enforcing energy efficiency? Why should the line be drawn at one point and not at another? Energy conservation did not

Figure 10.1   The major application routes

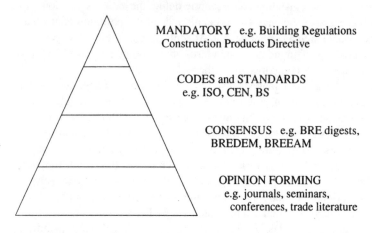

MANDATORY e.g. Building Regulations
Construction Products Directive

CODES and STANDARDS
e.g. ISO, CEN, BS

CONSENSUS e.g. BRE digests,
BREDEM, BREEAM

OPINION FORMING
e.g. journals, seminars,
conferences, trade literature

Source: provided by Neil Milbank, Building Research Establishment, 1993.

figure at all in the earliest building acts, which focused instead on aspects of design and construction directly affecting public safety. Buildings can be dangerous places and stability and fire proofing were of prime concern, particularly after the Great Fire of London (Knowles and Pitt, 1972; Muthesius, 1982). Although standards and expectations have changed considerably, arguments about health and safety continue to underpin the building regulations. Badly insulated buildings are not especially dangerous; they are no more prone to fire than any other, and while they might be uncomfortable to occupy they do not pose a threat to public health. So thermal regulation is exceptional in that it is justified in terms of the rather more slippery concepts of national interest, national fuel, security and, more recently, environmental protection. Arguments about thermal regulation are therefore arguments about national and/or environmental interest in energy efficiency. Other equally compelling areas of national interest – such as the health of the building industry and the cost of housing – complicate the equation, setting the scene for continual negotiation.

Adjustments in national and international priorities now mean that the government has an additional interest in changing the regulations in order to help meet targets for reducing carbon dioxide emissions. In this way, concern about 'the environment' has crept in, modifying the language employed by participants already engaged in long running negotiation. The vocabulary is likely to shift again, for EC Construction Products Directives relating to energy economy and heat retention will introduce yet another set of regulatory criteria. These are significant developments since there is no knowing where environmental concern might lead or what it might justify.

Arguments about health and safety are, it seems, easier to accept than those about global environmental change. Furthermore, concerns about health and safety are explored within a relatively finite arena defined by conventional perceptions of risk and by a familiar set of physiological requirements. The environment is not so bounded, and in any case, distant environmental anxieties do not present the same sorts of immediate, personal dangers as those posed by an unsafe building. There are signs that proposals designed to minimise environmental damage are already pushing at the limits of regulatory legitimacy: it is not clear that the environment should be protected in this way and, even accepting that it should, it is not clear how far the case should be taken.

'The environment' is thus placing increasing pressure upon a system initially set up to cope with the seemingly simpler tasks of making sure that buildings do not collapse or catch fire too easily. Any serious attempt to reduce carbon dioxide emissions will significantly extend the range of regulatory control, encompassing aspects of building design not previously subject to government surveillance (see, e.g., draft consultation documents relating to air conditioning and mechanical ventilation: DoE, 1993). More than that, such attempts are likely to be based upon new and unfamiliar measures (so many units of carbon dioxide per m^2 (Building Research Establishment Report No. 234, n.d.) which have no established meaning within the industry. Most remarkable of all, these developments, which threaten to modify the basis and the foundation of government intervention, are suggested in the name of something as intangible as global, not even national, environment protection. In political terms, the grounds for intervention have become more complex. Changes in the rationale for thermal regulation have in turn altered the nature of the debate, affecting both the substance of the argument and the evidence cited.

The two threads of conventional regulatory theory – the political and the technical – together sustain a view of policy and policy making as a somewhat abstract process. They reinforce the sense that decisions about how far, and how fast, we move toward higher standards of energy efficiency (or environmental protection, pollution control, waste management, air quality, etc.) depend on the strength of political determination and the extent of technological confidence.

Such an approach obscures the practicalities of regulation and draws a veil over the constraints and opportunities generated by the routine management of that process. Let us continue with the case of the building regulations.

Although driven by the powerful, if confusing, forces of national and environmental interest, building regulations rarely affect the existing stock. The assumption is that 'once built, a building will continue to satisfy requirements unless there is a "material alteration" or change of use' (Atkinson, 1993: 137). This is a significant, but perfectly understandable, limitation. If this assumption were not made, we would require an army of inspectors charged with the job of regularly reviewing every property in the country. Just imagine the time, cost, paper and energy involved in such an exercise.

Building regulations have another important limitation in that they focus almost exclusively on building structure and building fabric. This is in part because of the traditional emphasis on health and safety, plus an additional concern to control elements of the building which are especially difficult to modify if things go wrong (such as the foundations). Energy efficiency has been equated with insulation and standards have consequently focused upon the thermal performance of walls, roof and floor. There are other considerations too, for it makes little sense to try to regulate what cannot be monitored or controlled. Without the army referred to above it would be impossible to police the many minor modifications which people make during a building's lifetime, or to monitor the replacement of bits which rot, rust and simply wear out. As a result, current regulations have (as yet) nothing to say about the detailed specification of heating systems or of air conditioning and nor, of course, do they have anything to say about the way in which buildings are used. Both features make a huge difference to levels of energy consumption, yet both lie beyond the scope of government control.

Accepting the necessary focus on new building and on fabric and structure, is there anything to stop an environmentally enthusiastic government from turning up the regulatory dial as far as it will go, and achieving 'Scandinavian' standards overnight?

CONSTRUCTING REGULATIONS

Paradoxically, the fact that building regulations are mandatory in itself limits their power and influence. If builders and designers are obliged to adopt totally unfamiliar practices so many will fail to comply that the regulatory system will simply grind to a halt. As many have recognised, this means that regulations generally lag 'behind' current practice. There may, nonetheless, be some circumstances in which regulation forces the pace of technical change. It is not a clear cut question of the government dragging the building industry forward or, conversely, of the building industry holding the government back. What is at stake is a much more subtle balancing act in which the ambitions of the regulators are set against the current capacities of those they seek to regulate. But who is to say what those capacities really are?

The government is understandably unwilling to require builders and designers to do things which result in technical failure. This is tricky as buildings are complicated products, and measures which work well when properly applied can have disastrous consequences if handled badly or adopted in inappropriate circumstances. Assessments of the industry's capacity for technical change therefore involve assessments of the political risks of technical failure and of the equally political risks and costs of inaction. There is enormous potential here for argument, manipulation and manoeuvring and for trading between the parties involved in the debate.

It is important to recognise just how many parties are involved. Again the element of compulsion is a critical factor. Any change in the regulations has immediate knock-on consequences for those designing and constructing buildings and for those who produce and sell building materials. Whole livelihoods are at stake. Consider, for example, the practical consequences of the most recent edition of the thermal regulations. The few house builders who were not already specifying double glazing were able to meet the new standards by doing so: a response which has obvious consequences for the glazing industry. For reasons discussed below, the 1990 regulations did not necessarily mean that more insulation was sold, which has equally obvious consequences for the insulation industry. It is not just that there are many vested interests or that industry-based lobbying is an important part of the regulatory process; rather, it is the dynamic interaction of so many pressures which makes the consultation process so elaborate. Resulting tensions reflect the sheer variety of ingredients which go into making a building, and mirror perceived risks of technical failure and anxieties about additional cost.

Changes in the regulations, and even changes in the way they are implemented, make a difference to the sale of materials such as bricks, blocks, glass and insulation. With each change some companies gain while others lose out. Sectors dominated by a few large companies, or represented by particularly strong trade associations, have a better chance of influencing events than those which are less well organised. There is a sense, then, in which the institutional structure of the industry affects the regulatory process, also filtering associated cost implications. As noted above, it would not be in the national interest to demand standards of energy efficiency which were seriously 'uneconomic' and which endangered the commercial viability of the building industry. It is difficult to pin down these financial factors for costs are notoriously slippery. Currently costly items might become less expensive if everyone were obliged to buy them, and present rates are not an especially good guide to future prices. In any event, there is more balancing to be done, with costs of compliance weighed against associated environmental and/or national benefits. This is rather more than a technical accounting exercise. The meaning of 'uneconomic' is clearly not the same during a recession as it would be when house building and property development are booming. Furthermore, actual costs to builders – who engage in bulk buying deals and so on – are rarely the same as those predicted by building economists. While economic 'evidence' is as open to dispute as any other the real arguments tend to revolve around technical risk and potential as well as apparent financial implications.

Building performance depends on the interaction of many different elements. Tampering with any one part of the building may therefore have unforeseen (and possibly unpleasant) consequences elsewhere in the system. Not surprisingly, builders are reluctant to make what they see as potentially risky changes inviting untold financial costs. Some proposed changes to the building regulations are, of course, more worrying than others. There are, in effect, thresholds of anxiety within the industry. These thresholds define the scope for acceptable change at any one time. Accordingly, real controversies arises when thresholds are reached: that is, when the next step, in terms of energy efficiency, involves a major change in normal practice rather than a minor incremental adjustment. Current argument about fully filled cavities is a good example of just such a critical 'threshold' debate. Proposed amendments require an increase in the thermal performance of external walls which can 'only be easily obtained by full cavity fill', a strategy which meets with 'considerable resistance both by the builder and the house purchaser' (Beazer Engineering Services, 1993). Formal objections have flown from the House Builders' Federation and the National House Builders' (NHBC). Council. In this as in other instances, the interests of the NHBC, acting as the insurer of many private sector house builders, run counter to those of others within the building industry (such as the National Cavity Insulation Association), and indeed within the government.

Given its reliance on the scientific-technical theory of regulation it is perhaps not surprising to find that the government spent £8.1 million on building regulation research in 1990–1991 (DoE, 1992). Such research will, it is hoped, provide solid, unassailable, evidence about the costs, risks and anticipated environmental advantages of proposed changes in the regulatory system. Equipped with a body of unambiguous technical data, the government will then be able to steer its way through the fray of conflicting interests within the industry: or at least that is the hope. In reality, the heavy weight of research evidence does not necessarily ease the consultation process. It does, however, influence the nature of the debate, with competing interest groups then marshalling further, equally unassailable, evidence in support of their own claims. In this context it is important to recognise that perceived risks are far more important than 'real' risks established on the basis of survey data or experimental study. In their response to the proposed amendments, Beazer Engineering Services, on behalf of Beazer Homes and the House Builders' Federation, refer to the beliefs and faiths upon which current practice depends: 'It has always been one of the fundamental beliefs that a house should be able to breathe' (Beazer Engineering Services, 1993). Tangled and protracted negotiations ensue and it takes months, sometimes years, to reach the level of agreement on which effective regulation depends. Periods of recession sharpen differences within the industry, also modifying the alignment of interests and influence in the regulatory process. There is a sense, then, in which the detail of environmental regulation has as much, if not more, to do with the relative positions of the various actors within the building industry during the late 1980s and early 1990s than with the government's environmental policy or with the unstoppable march of technological progress.

We clearly need to review our theories of regulation. Neither the 'sledge-hammer' model nor the 'tip of the iceberg' theory of innovation adequately describe the essentially negotiated character of the regulatory relationship. Both overlook the role of the industry and the influence of the conflicting interests contained within it. While it would be wrong to cast the government as an innocent and somewhat stunned pig-in-the-middle, it is an image which captures at least some of the complexity described above. As we shall see, relationships which shape the construction of regulations also influence their practical implementation.

IMPLEMENTING REGULATIONS

It is tempting to forget implementation and assume that the story ends when new, environmentally improved, regulations are at last introduced. That temptation should be resisted since the day-to-day business of monitoring, testing and checking has important consequences for the people involved, for the efficacy of the legislation they police, and for the nature and character of the whole regulatory process.

Enforcement and monitoring is an essential part of regulation. Without it there would be no way of identifying those who break the rules and no knowledge of current practice. While there is clearly a need for some sort of inspection, there is plenty of scope for debate about the inspection process. The environment is already patrolled by a variety of different inspectors: some concentrating on water, some on air pollution, others on the built environment. In each context the ratio between the number of inspectors and the number of cases, people, sites, or applications to be reviewed differs. This is not necessarily a problem but it underlines the point that there is no self-evident definition of 'enough' inspection. It is not just a question of the number of inspectors; much also depends on what they do. There are any number of options here, each with different implications in terms of time and cost, and in terms of the scope and detail of the inspecting process. Should there be more site visits or less travelling and more telephoning? Are a handful of random checks more effective than an annual programme of announced inspections? How should samples be drawn, and what methods of measurement should be employed? Somehow these 'unanswerable' questions do get answered.

In the case of air pollution, 'A small staff of around 50 is responsible for inspecting 3000 separate processes at 2000 industrial sites. An average of 5 visits per year has recently been made, one-third less than a decade before' (Boehmer-Christiansen and Skea, 1991: 175). In the rather different case of cavity wall insulation, British Board of Agrément (BBA) approved installers are currently obliged to pay for two monitoring visits per year and one headquarters' inspection. In this instance, the level of inspection reflects a delicate balance between cost and rigour. 'Too much' inspection would place too great a burden on the installers who have to pay for their own monitoring. 'Too little' would undermine the perceived value of BBA approval, again creating problems for both sides. Definitions of 'enough' therefore reflect a 'gut feeling' of adequacy, loosely framed by a sense of value and an equally diffuse understanding of reasonable expense.

The cost of regulation also depends on the sort of measurement involved in assessing compliance and monitoring performance. In the case of building control, approval of drawings together with subsequent site inspection appear to cover all eventualities, but even here decisions have to be made about the practical meaning of checking, inspecting and monitoring compliance with the regulations. Building control officers check plans submitted for approval but do not always inspect the work itself. While that has remained the pattern for some time, the detailed process of checking for compliance has changed considerably. Some regulations specify an end result *and* the way in which that result should be achieved; others merely specify the end result. The latter strategy, which appears to have the advantage of ensuring control without unnecessarily restricting choice, informed the design of the 1990 thermal regulations. As a result, designers and builders are free to meet an overall energy-related performance specification in (more or less) any way they want. In the case of house building, more floor insulation but single glazing might, for instance, equate with double glazing and less floor insulation. Trade offs are possible. In this environment it is not enough to check over the details of wall, roof and floor construction: building control officers must now review the combined effect of all these elements in order to evaluate the performance of the system as a whole. Checking drawings for compliance with the regulations becomes more complicated. There is, of course, a point beyond which checking and inspection becomes so complicated that it does not serve its purpose as a practical, manageable method of regulatory control. After all, those who are being inspected must be able to understand and relate to the measures imposed upon them, just as those doing the measuring must be able to handle that process within accepted limits of time and cost.

Not surprisingly, there are some things which governments and others seek to control but which cannot be measured in these straightforward terms. Effective installation of cavity wall insulation is again a good example. Blowing foam, fibre or bead into an empty cavity wall is one of the single most energy efficient actions a householder can take (EEO, 1992), or at least it is if the job is done properly. The snag here is that short of an expensive (i.e., 'too' expensive) thermographic survey there is no way of telling if the job has been done well, badly, or even if it has been done at all (Shove, 1991). The BBA and the other assessing organisations in industry have no option but to rely upon proxy measures of technical competence and honesty. Surveillance therefore focuses on training procedures and administrative systems: efficient filing thereby becomes a measure of efficient filling. Methods of measurement have to work within confines of the monitoring bureaucracy. This sometimes means that the 'wrong thing' is measured. In other cases, the practicalities of monitoring determine the choice of measuring instrument. There are, for example, different ways of measuring air and water pollution. The chosen solution for the purposes of routine monitoring is only one of a number of options. To understand the measuring process – and the implementation, impact, and effect of regulation – we need to understand the organisational contexts within which some measurement options count as manageable and others do not.

To summarise, the routine business of monitoring compliance and ensuring enforcement depends on careful record keeping. It is, at heart, a matter of managing information. This introduces a new set of considerations, for the language of manageable measurement is not a perfect translation of the macro language of environmental policy, carbon dioxide targets and the rest. The fine detail of site visits, interpretation, interaction and measurement are industry specific. Such details reflect idiosyncratic inspecting histories and current, somewhat arbitrary, arrangements relating to the scale, form, cost, and organisation of the inspection process. Policy makers' access to the realities of environmental action is thus mediated by the normally invisible business of implementation and data collection.

Although invisible, the taken-for-granted practicalities of implementation influence the regulatory system in a number of ways. Routine working arrangements clearly inform the definition of manageable measurement so determining the scope for future regulation. Just as important, methods of monitoring and implementation generate data which are then used to assess the consequences of past legislation and to identify opportunities for further control.

REGULATING CONSTRUCTION

The notion that Britain could achieve 'Scandinavian standards' by simple adjustment of the regulatory dial now seems highly problematic. Such a proposition assumes that environmental policies can be made and implemented by governments acting alone, driven by the internal logic of their own scientific evidence. This is clearly not the case. Neither is it the case that the pure environmental ambitions of committed politicians are confounded, hindered and thwarted by reluctant and conservative industries. Rather, opportunities for regulation depend upon, and are shaped by, the necessary involvement of all sorts of different participants. (Guy and Shove, 1993)

That is not to suggest that central government has no part to play in the process or that environmental enthusiasm is irrelevant. Along the way there have been a number of identifiable and important macropolicy changes which have had a bearing on energy conservation in buildings during the late 1980s and early 1990s. The notion that amendments to the building regulations might be justified on grounds of environmental protection and the reduction of carbon dioxide emissions is one such development. European Construction Products Directives look set to change the terms of regulatory debate, while the introduction of performance specifications and of other more complex and more 'flexible' forms of regulation is perhaps even more significant. Although important, these contextual adjustments do not determine the detailed substance of the regulations.

To understand the changing content of environmental policy we need to attend to the gritty realities of micropolitical power within the building industry as well as to the more formal statements of government ambition. The recession, the number of new houses built and sold, the state of the insulation industry, the market share of lightweight block: these are critical parts of the regulatory equation. Similar factors count in other sectors too. The number of companies involved in different industries, the weight and power of the pressure groups within those industries, the culture of regulation and the practical meaning of manageable measurement together constitute the essential ingredients of environmental policy making.

CONCLUSION

Academic analyses of environmental politics tend to gloss over these local, industry specific characteristics and in doing so inadvertently sustain the 'sledge-hammer' theory of environmental regulation. In such debates regulation features as one amongst other measures which governments can employ in bringing about desired environmental change. In discussion of alternative policy options, regulation is generally presumed to be a blunt but effective instrument. It is a stick rather than a carrot. More than that, the scope of regulatory control is frequently taken to indicate the strength of political will. By contrast, this discussion suggests that regulations reveal as much about the relative negotiating powers of the various groups involved as they do about the determination of a single minded government. Whichever way we look at it, the sticks of environmental regulation are really rather bendy.

References

Atkinson, G. (1993), 'A View of the Regulator's Perspective', in F. Duffy, A. Laing and V. Crisp, *The Responsible Workplace* (London: Butterworth Architecture).
Beazer Engineering Services (1993), *Proposed Amendments to Part L of Building Regulations: A Review and Cost Analysis.*

Boehmer-Christiansen, S. and Skea, J. (1991), *Acid Politics: Environmental and Energy Policies in Britain and Germany* (London and New York: Belhaven Press).

Building Research Establishment Environment Assessment Method report No. 234 (n.d.), Breeam/New Offices.

DoE (1992), *Research Report, 1990–1991* (London: DoE).

DoE (1993), Revised draft – consultation documents relating to Approved Document L1, Part L, Conservation of Fuel and Power, Extracts relevant to Air Conditioning and Mechanical Ventilation, 30 November.

EEO (1992), *Insulating Your Home*, autumn edn. (London: HMSO).

Guy, S. and Shove, E. (1993), 'Leaping the Barriers', paper presented at the British Sociological Association Risk and Environment Study Group, York, December.

Knowles, C. and Pitt, P. (1972), *The History of Building Regulation in London* 1189–1972. (London: Architectural Press).

Muthesius, S. (1982), *The English Terraced House*. (New Haven and London: Yale University Press).

Polan, B. (1990), 'Flaws and Defects', *Guardian*, 26 January.

Shove, E. (1991), *Filling the Gap: The Social and Economic Structure of the Cavity Wall Insulation Industry*, report for the Building Research Establishment's Energy Conservation Support Unit.

11 UK Environmental Policy and Transport

Kenneth Button

INTRODUCTION

Transport is a major cause of environmental degradation. The problems are neither new nor simple in their impact but certainly it has been the advent of motorised transport which has led to many of the current problems, a point eloquently made twenty years ago by J. M. Thompson (1974) when he described transport as:

> an engineering industry carried on, not privately within the walls of a factory, but in public places where people are living, working, shopping and going about their daily business. The noise, smell, danger and other unpleasant features of large, fast-moving machinery are brought close to people, with potentially devastating consequences for the human environment.

As a result of these problems, and the public concerns raised by them, transport has traditionally been the subject of a considerable number of environmentally based regulations and controls over the years. Indeed, as far back as Roman times restrictions were put on the use of wagons in urban areas to contain traffic noise and reduce dust levels. The nature of both the problems associated with transport and their magnitude, however, have changed with time, and with this has come a shift in the forms and intensity of official policy. In particular, over the past twenty years the general perception of the problem has moved from concern with local, and especially urban, environmental effects such as noise, vibration and emission of pollutants such as lead and black smoke (see Independent Commission on Transport, 1974), to the contribution that transport makes to transboundary environmental problems, such as acid rain, and to global environmental change, and especially its share of carbon dioxide emissions.

The objective of this chapter is initially to provide some details of the exact nature of the environmental damage done by transport activities in the UK. It then provides an account of the types of policy measure which have been pursued by the UK authorities in the past and looks at how these have changed in recent years. Given the UK's membership of the EU it is important that these policies are seen in their wider, pan-European context and especially within the framework of the environmental policy of the EU. A key point regarding transport is that by definition, it is a mobile source of pollution, and consequently it is impossible to isolate UK policy from that of the wider European space. However, no policy approach can be perfect, and the paper concludes with some thoughts on the success to date of transport policy in the UK with regard to the way it fits within more general environmental policies.

THE NATURE OF THE PROBLEM

The environmental implications of transport are both substantial and diverse (Banister and Button, 1992; Button, 1993). Indeed, it is the very diversity of the impacts which often pose difficulties since policies designed to reduce the intensity of one problem can lead to a worsening of others; for example, rough road surfaces reduce accident risks but increase noise levels. Equally, no mode of mechanised transport is free from adverse environmental consequences although their natures and intensities often differ. Added to this, the impact of any particular transport mode can vary according to its location, the way it is operated and its time of use. Car noise, for example, is not a major problem in an industrial area during the day time but the same vehicle can cause serious disturbance during the night in residential parts of a city, especially if it is driven harshly.

The scale of the problems of transport have, of course, grown with the growth of the sector itself. Equally, however, while all transport creates environmental problems it is road transport which is generally seen as the most damaging mode, and one reason for this is its sheer success in attracting this new traffic. For example, nearly 90 per cent of passenger kilometres is done by car, with over 80 per cent of tonne kilometres of freight being done by road haulage. Car ownership has grown as part of this trend (and also as a cause of it), and this has brought about secondary environmental problems in the manufacture and disposal of the vehicles.

Table 11.1 offers some quantification of a number of the key recent environmental trends associated with transport use in the UK. These are illustrative in their own right, but perhaps a more useful way of looking at the environmental problems associated with transport, and also setting them within an analytical and policy framework, is to draw up a simple taxonomy reflecting the scale of their geographical and temporal incidence and to offer a few brief comments on their social and economic implications.

Table 11.1   Atmospheric pollutant emissions from transport in the United Kingdom

Pollutant	1983	1985	1987	1989	1991	Per cent of all emissions
Nitrogen oxide	1070	1200	1370	1610	1650	60%
Carbon monoxide	4400	4710	5300	6120	6120	90%
Volatile organic compounds	1140	1170	1270	1350	1300	49%
Lead	6.9	6.5	3.0	2.6	2.0	100%
Black smoke	130	147	172	204	214	43%
Carbon dioxide	105200	111200	117200	129300	138300	24%

Local Effects

First, transport imposes a wide range of what can be called local effects on those people living and working in areas close to the transport activity. These local impacts are themselves diverse and range from noise nuisance through community severance, visual intrusion and accident risk, to emissions of a variety of toxic gases. Many of the consequences of these local environmental effects can be quite damaging to health, with several of the local pollutants being associated with heart disease and cancer. Others, such as noise, are primarily disruptive in their nature but can still produce ill-health. In general, however, these effects are short lived in the sense that the removal of traffic or a change in technology will very quickly reverse the problem. They seldom have cumulative effects. Equally, it may be possible for individuals to avoid many of these problems or to take actions (such as double glazing against traffic noise) which reduce their adverse social consequences.

It has been the set of local environmental transport induced effects which has traditionally attracted public interest and policy response. Traffic noise in cities has, as noted earlier, been the subject of public concern at least as far back as Roman times, and emissions of pollutants, such as lead, have resulted in widespread popular calls for policy responses. Fortunately, and unlike parts of countries such as Greece and parts of the west coast of the USA, some of the more severe problems of local tropospheric smog brought on by processes of photochemical reactions of gases are much less pronounced in the UK (although such low level ozone problems can, and sometimes do, arise in certain weather conditions).

Transboundary Effects

Second, there are the effects of transport which cross administrative boundaries or borders and where the environmental damage done is some distance from the transport source itself. Emissions of nitrogen oxide, sulphur dioxide and other gases which contribute to acid rain are perhaps the most obvious examples. The outcome is that forests and fresh water reservoirs can be damaged and polluted. One can add problems of oil spillage at sea and also, possibly, the run-off from roads into local water courses, both of which often have impacts well away from the transport itself. The total annual yields run off from a kilometre of heavily trafficked rural motor ways, for instance, amounts to 1500 kg of suspended solids, 4 kg of lead, 125 kg of oil and 18g of polynuclear aromatic hydrocarbons.

Also, there is the matter of the materials used in the construction of vehicles and transport infrastructure (including the energy used in their production as well as the raw materials which are consumed) which can have major implications vast distances from where the vehicles are used or infrastructure provided. This is not only a problem for land transport, because considerable quantities of materials are dredged to keep maritime channels open and this is often dumped in open waters.

Global Effects

Finally, there are global issues which have transcontinental implications. These embrace high levels of carbon dioxide and other GHG emissions which have associated with them possible implications for global warming (Hughes, 1993), and activities, such as the high flying of commercial aircraft, which may influence the depletion of the stratospheric ozone level. The former has potential implications for the thermal expansion of the sea, while the latter's depletive effect on higher ozone cover can result in cancers. Although transport in the UK is not the major contributor to carbon dioxide emissions (it accounts for about 24 per cent of the total) it is the fastest growing sector. Given the nature of the problem it must also be taken in a broader geographical context. If UK transport was the only source of the increased carbon dioxide pollution then the problem would be minimal, but in fact a similar pattern of carbon dioxide increase is being observed across Europe (Whitelegg 1993) and elsewhere.

It should also be remembered that transport infrastructure covers a large amount of land (for instance, between 1.2 and 1.5 per cent of the UK land area is covered by roads) which may, by disrupting and destroying natural habitats, be contributing to reductions in biodiversity. Finally, there is the question of the depletion of scarce mineral resources. Strictly, this is not a problem if depletion is at an optimal rate with prices for these resources reflecting their true opportunity costs. In practice, it seems unlikely that such prices are currently being levied and depletion may, therefore, be excessively rapid. Such resources include a variety of metals, but transport is also a major user of energy in the UK and, in particular, fossil fuels (see Table 11.2).

Trying to put these diverse effects into a broader, economic context is not easy. It is, however, important to do so if sensible trade offs are to be made not only between actions to contain the environmental damage done by transport *vis-à-vis* other sectors but also between the diverse effects of transport itself. What this means in practice is that policy often requires trade offs to be made. The removal of lead from petrol to prevent potential brain damage to children, for example, may reduce the harm done by that pollutant but only at the expense of increasing other local atmospheric pollutants, such as benzene (a carcinogen), or at the cost of higher fuel consumption with consequences for carbon dioxide emissions and the depletion of a non-renewable resource.

In an effort to try to highlight the nature of the trade offs which are often necessary, efforts have been made to put a monetary valuation on some of the local environmental implications of transport and, indeed, a recent report of the Standing Advisory Committee on Truck Road Assessment (DoT, 1992) has suggested experimentation with putting environmental costs into the current methodology for appraising road infrastructure investments. At a more macro level, Pearce (1994) has attempted to place an aggregate value on the overall social costs of road transport in the UK, which embraces congestion as well as environmental effects, and concludes that they amounted to between £22.9 and £25.7 billion in 1991, which was about twice the revenue the Exchequer took from road users. While one can easily question the details of Pearce's arithmetic – and, indeed, the underlying logic of trying to aggregate values often derived from partial equilibrium studies may be considered suspect – there is mounting evidence that the net environmental costs of transport are extremely large.

Table 11.2   Energy use by transport

Year	1960	1965	1970	1975	1980	1985	1987
Percent of UK energy used by transport	17	18	19	22	25	27	29

Source: UK Department of Energy,1988.

The question of why these problems of excessive environmental degradation arise from transport has traditionally been thought of in terms of market failures. Efficient resource utilisation comes about if appropriate markets are defined and they are allowed to operate freely. The problem with transport is not only the conventional one encountered in virtually all sectors (namely, that there is often no effective market for environmental goods), but also the fact that transport itself is often not provided by efficient market mechanisms.

For a diversity of reasons much transport infrastructure, and most notably road capacity, is provided on political criteria and no direct charge is made for its use; in virtually all industrialised countries roads are treated as a merit good with fuel taxation and annual excise duties seen as sumptuary taxation. There are also large subsidies given for transport use, and transport activities are the subject of a plethora of regulations. Since many of the distortions are not intrinsic to the nature of the transport market, these problems are often much more to do with government intervention failures than with strict market imperfections, as the case studies in Barde and Button (1990) have illustrated. Without wishing to preempt subsequent discussions, this leads on to some interesting questions as to the extent to which further interventions could simply worsen the problem.

EMERGING TRANSPORT TRENDS

Transport poses an environmental problem at current levels of use but the indications are that if previous trends continue without remedial action these problems will get worse. In particular the environmental damage done by transport is highly correlated with transport growth itself. Land passenger travel (other than walking and cycling), for instance, increased from 219 billion passenger kilometres in 1952 to 681 billion in 1993. Further, the 42 per cent growth in UK road traffic between 1980 and 1990 was the fastest in Europe. Road transport – which, as we have seen, now accounts for about 90 per cent of passenger traffic – is a major polluter and the official forecasts for the UK (Table 11.3) are that this will continue to grow at a rapid into the next century. Equally, aviation has grown rapidly over the past decade and the number of movements through the South-East of England alone is projected to rise from 69 million in 1992 to 75 million by 2000 if current policies are continued.

While part of this road traffic growth can be accounted for in terms of additional trips, this represents a relatively small part of the recent and predicted increase. For a variety of reasons, including the relatively low cost of road use, rising incomes and changing consumption patterns, a

Table 11.3   Forecasts of motor traffic on all British roads (excluding motor cycles)

Year	Low		High	
	bn vehicle km	*index*	*bn vehicle km*	*index*
1990	346	105	358	109
1995	383	117	419	128
2000	418	127	482	147
2005	453	138	544	166
2010	488	149	606	185
2020	559	170	728	222
2025	595	183	789	242

Source: UK Department of Transport, 1989a

more important factor is the increased length of average trips (Table 11.4). While traffic growth in general has serious implications for the environment, this pattern of travel behaviour is leading to increased degradation of the countryside because these longer journeys are associated with suburban living, the extension of urban sprawl and leisure trips into areas of natural beauty. These figures also exclude the growth in travel to more distant overseas destinations for holidays.

At the global level a number of important trends will inevitably lead to growing concerns about carbon dioxide and other GHG emissions. Emissions of these gases are highly correlated with carbon fuel use which, in turn, is highly correlated to the level of traffic. Globally, the freeing-up of international trade made possible by recent General Agreement on Tariffs and Trade (GATT) agreements and the creation of large economic unions (such as the European Single Market and the North American Free

Table 11.4   Average personal car travel in the UK

	Journeys	*Miles per person a year*
1972/73	956	4,460
1975/76	935	4,710
1978/79	1,097	4,950
1985/6	1,025	5,320
1989/91	1,090	6,480

Trade Agreement) will stimulate transport growth. In the post-Communist countries of Eastern and Central Europe, as economic systems are restructured and consumption levels rise so the demands of consumers and manufacturing industry for transport services will grow. Equally, there is evidence that in many of the traditionally poorer countries of the world rising incomes and higher consumption levels will bring forth demands for higher car ownership levels (Button, Ngoe and Hine, 1993). All these factors will produce higher levels of global carbon dioxide emissions and, unless the currently affluent countries in some way suppress the aspirations of these emergent economies, there will be a need for them to reduce their own emissions of GHGs considerably.

THE DEVELOPMENT OF UK POLICY

Traditionally, UK transport policy has focused on local issues and, in particular, the urban impacts it has on the environment. Public concern over noise and the immediate physical impacts of transport, such as diesel fumes and safety have been at the root of much public policy. Indeed, noise is often recorded in surveys as one of the worst aspects of urban living. The policy response, until comparatively recently, has centred on planning and technical solutions. Often these have combined efforts to link the environmental elements of policy with those designed to improved transport services for users.

Such policies have often failed, however, to achieve their stated objectives fully. Enforcement of controls (such as overparking) has often been a problem, but the difficulty has mainly been inadequacies in the traffic restraint elements of the policy packages and, in particular, an underestimation of the public's desire to own and use cars. These types of measures should not, however, be altogether underestimated in their importance. Simple regulations such as legal requirements to drive on the left, maximum speed limits, one-way streets, parking restrictions and the allocation of junction priorities generally help to smooth out traffic flows, and this reduces noise and emissions of most types.

More explicit in the post-Second World War period have been the physical planning efforts which can most directly be traced back to the Buchanan Report (UK Ministry of Transport 1963) on *Traffic in Towns*, although there were several antecedents. The key belief at the time was that it is possible to plan urban areas in such a way that the local environment can be protected while still affording high levels of personal mobility, particularly to the private motor vehicle. Indeed, in a sense this philosophy still lingers on, albeit much less so at the urban level, with recent road planning statements. A substantial part of future planned investment is aimed at the construction of nearly 1100 kilometres of by-passes (DoT 1989b) as an element of a policy designed to separate through road traffic from sensitive parts of urban areas. Green Belt strategy, while not explicitly transport driven, follows a similar vein since it removes many of the attractions for non-leisure movements in these areas.

At a more micro level, the planning and design of transport infrastructure itself was recognised in the late 1980s as an important tool in environmental policy. In particular, and following the example of a number of continental European countries, measures such as traffic calming (the use of such things as speed ramps, raised junctions and reduced carriageway widths) to slow urban traffic and to encourage the use of 'suitable' links were initiated. While the focus in the UK has been on reducing accidents – about 70 per cent have this as their objective – they also have been seen as part of wider environmental packages.

Coupled with this planning strategy has been a tradition of trying to attract transport users to less environmentally intrusive modes of transport and, especially, to try to attract users away from privates cars and freight consigners from road haulage. Carrots have regularly been offered in the form of public transport subsidies or investment finance (e.g., for private rail freight sidings) to encourage such a switch. This has been supplemented, as under the 1968 Transport Act, by measures designed to improve the quality of urban public transport in particular by enhancing the mechanisms for ensuring 'coordinated services' which would be attractive to potential users. The empirical evidence, however, suggests that measures of this type are not, especially in the short term, in themselves sufficient to attract large numbers of people to switch from their cars (Goodwin, 1992).

Increasingly complex consultation procedures have also been developed aimed at ensuring that environmental factors are brought into land-use planning and infrastructure investment decision making. A major problem with transport investment decisions is that while fairly comprehensive computerised models, such as COBA, are available to assess the purely transport costs and benefits of investment options (most notably travel time and vehicle operating cost effects), most third party environmental effects have been handled separately, usually within an environmental impact statement. This has not only meant that decisions have been based on comparisons employing different presentation devices but also, and this is of increasing concern, usually ignoring all but the most localised environmental repercussions.

Vehicles themselves have traditionally been heavily regulated, especially for safety reasons. The need for older vehicles to pass annual Ministry of

Transport tests for road worthiness, while now serving an explicit environmental as well as safety purpose, has always implicitly helped contain many environmental problems. There have also been long standing laws on noise levels for virtually all modes of transport, and these have gradually been tightened in recent years. Perhaps the most important of the policies in the 1980s was, following EC debates, the decision that all new cars from 1992 must be fitted with catalytic converters. Catalytic converters reduce both nitrous oxide, hydrocarbon and carbon monoxide emissions per vehicle by about 80 per cent, although they are only fully efficient once engines are warmed up and they do increase energy consumption and thus carbon dioxide emissions. Whether this end-of-pipeline technology is preferable to other technical approaches (such as the lean-burn engine which has the potential to reduce nitrous oxide and carbon dioxide emissions) has been a point of discussion, and may be seen as representing some of the problems of initiating technological policies where much of the debate is captured by vested interests, such as those of large manufacturers.

Fiscal instruments, other than subsidies, have, however, historically formed only a very small element of UK policy. Annual road user charges (fuel tax and annual excise licence) in the UK are regularly calculated by the DoT to exceed expenditures on new road construction and network maintenance costs by over 2.5:1. One of the justifications for this has traditionally been that this, at least in part, reflects some of the social costs of transport. The lack of any direct linkage, however, either between the form of taxation adopted and the nature of the environmental damage done or between the categories of vehicle using roads and the specific polluting characteristics, does suggest that sumptuary taxation is a more likely reason for the taxation-expenditure divergence. More recently, tax differentials favouring unleaded petrol have, in combination with the statutory removal of low grade leaded petrol to free up pump space and a gradual reduction of the amount of lead permitted in leaded fuel, resulted in a significant fall in the level of inorganic lead pollution in the UK (see Table 11.1). Finally, one should also mention that, in the aviation sector, a number of major UK airports have differential landing fees for aircraft based upon the noise nuisance which they cause.

THE CURRENT UK APPROACH

A combination of domestic and international forces have brought about quite important changes in UK transport policy in the 1990s. First, the high level of traffic growth forecast for roads (Table 11.3), but also for other modes such as aviation, is unlikely to be acceptable either from a local environmental perspective or from the point of view of funding the increased infrastructure investments required to cope with it. Linked with this is a general change in attitude with respect to the way resources are used within the economy and an increased emphasis on efficiency. The inability of infrastructure to expand to meet unrestrained demand also seems inevitable even if more private financing of transport infrastructure is, as proposed, forthcoming (DoT, 1993).

Effectively a sea-change has occurred in policy, shifting away from the traditional strategy of meeting demand by expanding the capacity of the system, and towards one of managing demand within the bounds of an acceptable infrastructure network. This inevitably means greater use of traffic restraint measures and (although this is more a policy of pragmatic acceptance of the usefulness of the instruments) more widespread deployment of fiscal tools.

Second, the UK is part of a wider economic and political area and the priorities of the other EU member states inevitably influence what the UK can do or, indeed, must do. The Union has developed a strategy on transport and the environment (EC Commission 1992, 1993) designed to be compatible with sustainable development. The focus of what are still, essentially, broad policy statements is that traffic management is important for environmental improvement and that the planned development of various pan-European networks should be taken within this context.

Third, public perceptions of the nature of the environmental problems associated with transport have shifted. Certainly local issues are still central to the debate, although they now often tend to focus on protecting sites of historical or natural beauty rather than narrower concerns over noise and vibration, but increased information has brought the wider global issues to public attention. The incentive to act, especially in concert with other countries, on these wider environmental costs is enhanced by the similar nature of the problems they are being confronted by and the equal strength of concern being felt in most other countries, and especially in those with which the UK competes (OECD, 1988). Table 11.5 provides some supporting evidence for this. Effectively, acting to contain the problem simply shifts the basis of industrial competition for all involved and, while there may be efforts to enjoy free rider status, in reality the economic burdens will to a large extent be shared.

This latter sharing process has manifested itself in the numerous internal agreements which have been arrived at in recent years, such as the Montreal Protocol on the ozone layer depletion and the Earth Summit's Framework Convention on Climate Change. A general theme which

Table 11.5   Environmental impacts of transport across industrial countries

	North America	OECD Europe	Japan	OECD States
	Air			
	Total transport emissions as % of total emissions			
Nitrogen oxides (NO$_x$)	47%	51%	39%	48%
Carbon monoxide (CO)	71%	81%	na	75%
Sulphur oxides(SO$_x$)	4%	3%	9%	3%
Particulates	14%	8%	na	13%
Hydrocarbons (HC)	39%	45%	na	40%
	Noise			
	Population exposed to road traffic noise over 65dBA			
	19 million	53 million	36 million	110 million

emerges is the policy of setting general targets, with governments free to meet them in the way they deem most efficient. The UK approach in the transport sphere, and one which ties in with other aspects of both transport and environmental policy, has been to explore the scope for deploying more market based instruments. These major new fiscal policy responses have been on three main fronts.

First, and directly in response to the requirements for the UK to conform to the agreement reached at the Rio Earth Summit, increased fuel duty of, on average, 5 per cent per annum (pushing the real price of fuel slightly above their 1980 historical high point) has been initiated to help meet the Rio target of reducing carbon dioxide emissions to 1990 levels by the year 2000. The use of the fiscal instrument, while not the only policy adopted, nevertheless represents a major component in the UK's strategy in this field (DoE, 1994).

Second, proposals have been made (DoT, 1993), and supported in the second 1993 budget, to introduce electronic tolling for motorways which will act, partly, to contain traffic growth.

Finally, the government is funding major research into the possibility of introducing road pricing into large cities such as London with the aim of reducing traffic congestion and fostering the use of public transport.

The underlying philosophy of the approach is to leave transport users (and potential transport users) to decide on their own personal response when confronted with a transport costs structure which more closely reflects the resources involved. In that sense it may be seen as focusing as much on removing intervention failures as on reducing market imperfections. A difficulty, however, with all these measures and proposed measures, is to separate the environmental objectives of the policies from other aspects. In the case of the fuel tax, for instance, this is being introduced during a period of recovery from recession when containment of public expenditure, coupled with a gradual dampening down of economic expansion, form key components of macroeconomic strategy. Equally, the proposed motorway tolls (suggested as likely to be about 0.5p a mile for cars and 1.5p for goods vehicles) may be taken as simply a means of speeding up road construction and maintenance. Road pricing in urban areas, while likely to have positive environmental repercussions, can also be treated as a new source of local government finance. Hence, while the fiscal package may be seen as a cornerstone of recent UK policy, its exact role is still rather murky.

There are also a number of further issues which need resolution. In particular, the fiscal policy measures, while welcomed because they tend to have an impact on some of the root causes of the transport problem rather than just treating the symptoms, will, even if fully implemented, take time to any major effect on the situation. One is essentially dealing with comparative statics where the current situation is being compared to one ten years or so hence, when fiscal measures have had the chance to work. Mechanisms to adjust prices over time as traffic conditions change, or to introduce supplementary measures (such as more short term investment funding for public transport in cities as road pricing reduces car trips) to facilitate a smoother transition, are lacking. In some ways intervention may be counter-productive in that it could lead to wrong signals about the long term intention of pursuing the fiscal approach, but equally some balance needs to be struck between that and possible political rejection of the overall framework.

While national policy is important for global and transborder issues, there is a perceived need for more localised actions both to support the national strategy and to confront local environmental problems. The emerging roles of the UK central government in this context is that of facilitator, in the sense of giving local authorities the power to pursue their own initiatives and providing them with finance for implementation; of coordinator, so that authorities neither pursue beggar-thy-neighbour strategies nor adopt local policies which conflict with national objectives; and of educator, in that guidance and advice is available. The exact details of how these roles are to be fulfilled are yet to be made clear. Some guidance on appropriate planning of local land use are being devised to reduce travel needs, and strategies to make more productive use of derelict urban land are in hand. Much, however, seems likely to rest upon the success of devising ways of restraining car use and, in particular, whether measures such as road pricing can be made operational.

CONCLUSION

The diverse nature of the environmental costs of transport, coupled with the very mobility of the source of much of the pollution, has made it difficult for most governments to come up with a comprehensive environmental strategy. The continual rise in the demand for personal mobility, combined with the changing nature of industrial demands for transport, and especially with increasing emphasis on more direct control over movements, have led to UK society becoming highly dependent on roads for domestic transport. While all forms of motorised transport have associated with them a cocktail of adverse environmental effects, those associated with large scale road transport pose specific difficulties for policy makers.

The UK policy in the past has really represented one of accommodation of road transport with efforts to meet minimum standards with respect to local environmental quality. In terms of the criteria of the time, these policies can be seen as often being successful in meeting the specific targets set. Lead pollution, for example, has been reduced considerably; traffic safety is now better than at any time in the last thirty years; and noise levels have been contained in many sensitive parts of urban areas. Whether these trends can be expected to continue, though, in the face of predictions of high levels of future traffic growth can be questioned. There is also the problem that, in many cases, these policies have themselves produced their own problems: road building, for example, may have smoothed traffic flows with environmental benefits in the short term but has then itself generated more traffic, and while taking traffic away from sensitive areas has often attracted subsequent residential developments along its track.

The scale and nature of the problems have also clearly changed, both in terms of public perception and scientific emphasis, and have brought into question the ability of many of the old policy instruments to be effective in the new situation. The transborder and global nature of these concerns have inevitably led to them being looked at in continental or world terms and requiring the UK to fit in with wider policy initiatives. These facts have called forth a new approach to containing the environmental costs of transport in the UK. Greater emphasis is now being placed on the role of fiscal instruments both directly, to tackle some of the environmental impacts, and indirectly to reduce the growth in traffic volumes. The overall picture is not, however, complete at the time of writing. In particular, the role and powers of subnational authorities in meeting national policy aims are still being defined.

References

Banister, D. and Button, K. J. (eds) (1992), *Environmental Policy and Transport* (London: Spon).
Barde, J. P. and Button, K. J. (1990), *Transport Policy and the Environment: Six Case Studies* (London: Earthscan).
Button, K. J. (1993), *Transport: The Environment and Economic Policy* (Aldershot: Edward Elgar).
Button, K. J., Ngoe, N. and Hine, J. (1993), 'Modelling Vehicle Ownership and Use in Low Income Countries', *Journal of Transport Economics and Policy*, 27, 51–67.
Department of Energy (1988), *Digest of UK Energy Statistics* (London: HMSO).
DoE, (1994), *Sustainable Development – The UK Strategy* (London: HMSO) Cmnd 2426.
DoT (annual), *The Allocation of Road Track Costs* (London: DoT).
DoT (1989a), *National Road Traffic Forecasts (Great Britain) 1989* (London: HMSO).
DoT (1989b), *Roads for Prosperity* (London: HMSO) Cmnd 693.
DoT (1992), *Assessing the Environmental Impact of Road Schemes* (London: HMSO).
DoT (1993), *Paying for Better Motorways* (London: HMSO) Cmnd 2200.
EC Commission (1992), *The Impact of Transport on the Environment – A Community Strategy for Sustainable Mobility* (Luxembourg: European Commission).
EC Commission (1993), *The Future of the Common Transport Policy: A Global Approach to the Construction of a Community Framework for Sustainable Development* (Luxembourg: European Commission).
Goodwin, P. B. (1992), 'A Review of New Demand Elasticities with Special Reference to Short and Long Run Effects of Price Change', *Journal of Transport Economics and Policy*, 26, 155–70.
Hughes, P. (1993), *Personal Transport and the Greenhouse Effect* (London: Earthscan).
Independent Commission on Transport (1974), *Changing Directions* (London: Coronet).
Ministry of Transport (1963), *Traffic in Towns* (Buchanan Report) (London: HMSO).

OECD (1988), *Transport and the Environment* (Paris: OECD).

Pearce, D. W. (1994), *Blueprint 3 – Measuring Sustainable Development* (London: Earthscan).

Thompson, J. M. (1974), *Modern Transport Economics* (Harmondsworth: Penguin).

Whitelegg, J. (1993), *Transport for a Sustainable Future: The Case for Europe* (London: Belhaven Press).

12 Acid Rain: A Business-as-Usual Scenario

Jim Skea[1]

INTRODUCTION

Acid rain dominated the environmental policy agenda during much of the 1980s and, for the UK, it became the defining issue of the decade. Environmental groups dubbed the UK the 'dirty man of Europe' because of its reluctance to participate in the '30 per cent club' agreement to reduce acid emissions. The position on acid rain came to symbolise what many saw as a wider weakness in environmental policy making. The agreement of the EC's LCPD in 1988 signalled the end of an era marked by debate about whether and by how much emissions of sulphur dioxide and nitrogen oxides, the main precursors of acid rain, should be cut.

However, the acid rain problem has not been 'solved', and neither has discussion about measures to reduce acid emissions abated. The issue has moved from a political to an administrative mode where commercial and bureaucratic 'insiders' negotiate over the *means* of reducing emissions and the consequences of the regulatory methods used. For the electricity generation sector, which is responsible for a large proportion of UK acid emissions, acid rain remains the most important environmental issue influencing investment decisions. The regulatory and technical means to achieve emissions objectives remain unresolved, while the undemanding nature of current plans and obligations leaves an unfinished agenda in terms of international commitments. -

The chapter proceeds as follows. First, the key actors for acid rain policy are identified and their interests analysed. Their positions on acid rain are aligned with three different approaches to policy making: science based; cost benefit based; and technology based. The chapter then analyses the development of UK policy in three phases, assessing whether the nature of the underlying forces has changed over time or whether more persistent themes can be identified. The three phases cover the periods: pre-1988; 1988–93; and from 1994 onwards. The chapter concludes that the historic pattern of a clash between external pressures and domestic market/commercial interests is likely to continue.

The year 1988 is taken as the 'watershed' for acid rain policy. Not only was this the year of the then Prime Minister's famous Royal Society speech, it also usefully marks the agreement of the EC LCPD, the start of the electricity privatisation process and the point at which significant changes in the UK pollution control regime began to emerge. These developments were in fact more significant than the 'greening' of government.

UK POLICY ON ACID RAIN: THE ACTORS

Overview

UK acid rain policy has been formed by a small network of institutional actors. By dubbing the UK the 'dirty man of Europe' in the 1980s, NGOs were successful in raising public consciousness through a high profile *Stop Acid Rain* campaign. However, they have had a remarkably low influence over policy.

The debate over acid rain and its control may be thought of as following three parallel courses, as described below.

1.   *Science based,* driven by the need to achieve demonstrated ecological goals. The use of the critical loads concept, described below, exemplifies this approach which is implicitly adopted by NGOs.
2.   *Cost benefit based*, which is concerned with the appropriate balance between the costs of environmental damage and the costs of pollution abatement.
3.   *Technology based*, which focuses on the application of technology to reduce emissions. The BAT principle embodies this type of approach.

Table 12.1 identifies the key actors relevant to UK acid rain policy and identifies which type of discourse they have tended to adopt.

While there may be 'phantom constituencies'[2] which consistently support the use of one or other of these approaches, organisations with bureaucratic or commercial interests at stake tend to be more flexible in aligning themselves with any particular set of concepts.

Principal Actors

Department of the Environment

The DoE has the main responsibility for acid rain policy. It has pressed for more ambitious policies but its position within Government is relatively

Table 12.1   Actors in acid rain policy making area

Actors	Organising principles
Principal	
DoE	Primarily science based; since 1989 also interested in environmental economics ideas
DTI	Cost-benefit/economic efficiency
Electricity supply industry	Initially science based; now economic efficiency; but very low public profile on acid rain
Secondary	
HMIP	Technology based
The Treasury	Cost-benefit
Manufacturing industry	Cost-benefit/economic efficiency

weak. The political significance of broad public support for environmental issues is the strongest card which it has to play. The DoE's informational strength lies with the *science* of acid rain. It is weaker on issues relating to technology and emission abatement costs. In 1989–90, the DoE championed the environmental economics approach (DoE, 1990a) and, during 1994, a set of consultations will take place to assess the possible use of sulphur dioxide emission trading within the UK (DoE, 1993).

Department of Trade and Industry

The DTI and the Department of Energy acted as 'sponsors' within the government system for the nationalised electricity industry. Although the commonality of interest between the DTI and the electricity generators has diminished since privatisation, the DTI has continued to have an interest because: (a) the Director of Electricity Regulation reports to it; (b) the Government until recently retained a 40 per cent share in the two largest generating companies; and (c) the coal industry remained under national ownership. The DTI retains a strong internal capacity for assessing the economic implications of emissions abatement.

Electricity Generators

Expensive acid rain controls divert managers away from their core commercial concerns, eat up capital resources and reduce profitability. This applies equally to the Central Electricity Generating Board (CEGB) and its privately owned successors. Since privatisation, National Power and

PowerGen recognise that they must inevitably lose electricity market share and consequently their strategy is to diversify out of their traditional market. The funding for diversification comes from current profits in the core generation business. Expensive environmental controls would damage their capacity to achieve broader strategic goals. Under national ownership, the CEGB could justify a major programme of scientific research on acid rain in the wider public interest. The results were used to argue against the need for acid emission controls. In the early 1980s, the CEGB's expertise in relation to acid emission abatement was very limited. However, the industry acquired knowledge quite rapidly. The generators' strengths in acid rain discussions lie with DTI support and the very significant informational advantages regarding technology and economics which they enjoy with respect to HMIP, the regulatory body.

Secondary Actors

Regulatory Bodies

HMIP is formally responsible for pollution control at major industrial sources, including power stations, using the BATNEEC principle. However, it does not play a significant role in policy making.[3] Electricity privatisation has made a more formal relationship between Government and the electricity supply industry necessary and HMIP plays a bridging role by developing and applying 'regulations' which set the framework for emission controls at individual power stations. Although there is more knowledge about clean-up technologies within HMIP than in any other part of the government system, it still suffers from severe informational and analytical disadvantages when dealing with sophisticated plant operators such as those in the electricity sector.

The Treasury

The Treasury has had a vital, albeit background, role to play in relation to the acid rain issue. It effectively held a veto over the CEGB's pollution control expenditure which was paid for from public sector finances. While the DTI and the DoE were divided, this veto was exercised. When, as in 1986, the DTI and the DoE agreed over emissions controls, the Treasury was not in a position to object.

Other Industrial Interests

Although the electricity supply industry is the dominant source of acid emissions, other sectors also contribute. Both petroleum products and the petroleum refining process are a significant source of sulphur dioxide emissions. Controls on refinery emissions have not been strict, probably because of significant autonomous reductions in emissions in the early 1980s and because the complexity of refining processes has proved a barrier to regulators. Refiners face capital expenditure because of a recent EC Directive on the sulphur content of gas oil.

Manufacturing industry took a significant interest in the acid rain issue in the early 1980s, but mainly because of concerns about the impact of controls on electricity prices. Overall, manufacturing industry has played little part in the development of acid rain policy.

THE LEGACY OF THE 1980s

International Obligations

The UK is bound by the UN Economic Commission for Europe's (UNECE's) Convention on Long Range Transboundary Air Pollution (LRTAP) which was agreed in 1979 (Wetstone and Rosencranz, 1983). This requires signatories:

(a) 'to endeavour to limit and, as far as possible, gradually reduce and prevent air pollution'; and
(b) to use the 'best available control technology economically feasible'.

In 1985, the Convention was followed by a specific Protocol which requires a 30 per cent reduction in sulphur dioxide emissions from a 1980 baseline by 1993. The UK did not sign, arguing that scientific evidence did not justify the costs which would be entailed. Ironically, the UK will more than achieve the 30 per cent goal although very few specific emission reduction measures have been taken (see Table 12.2). This is due to recent investment in combined cycle gas turbine (CCGT) stations and improvements in the availability of nuclear plant.

The UK subsequently signed protocols covering emissions of nitrous oxide in 1988 and volatile organic compounds (VOCs) in 1991. The nitrous oxide protocol requires a stabilisation of emissions at the 1987 level by 1994. The VOC protocol requires a 30 per cent reduction in emissions by 1999 with respect to 1988 levels. The introduction of vehicle emission controls will assist in meeting the requirements of these protocols.

Within the EC, the UK agreed to modest phased reductions in sulphur dioxide and nitrous oxide emissions from large combustion plant under

Table 12.2   UK sulphur dioxide emissions (ktonnes/year)

	1970	1980	1983	1987	1991
Large Plant	4424	3867	3125	3264	2973
Power stations	2913	3007	2631	2830	2534
Refineries	213	237	117	102	115
Other industry	1298	623	377	332	324
Small Plant					
Industry	980	470	285	251	245
Other	1020	561	451	383	347
Total	6424	4898	3861	3898	3565

Source: Digest of Environmental Protection and Water Statistics.

the 1988 LCPD (Boehmer-Christiansen and Skea, 1991). The targets set were, for sulphur dioxide, 20 per cent by 1993; 40 per cent by 1998 and 60 per cent by 2003. The nitrous oxide reduction targets were 15 per cent by 1993 and 30 per cent by 1998. The LCPD also set strict emission limits for new power stations. The LCPD is a greatly watered down version of the original proposal made by the European Commission in late 1983. This would have required the UK to reduce its sulphur dioxide emissions from existing large combustion plant by 60 per cent by 1995.

Domestic Implementation

In 1986, the CEGB volunteered to retrofit expensive FGD equipment to 6000 MW of coal-fired power stations following an examination by its Chairman, Lord Marshall, of preliminary evidence from Scandinavian acid rain projects. The LCPD was then agreed in June 1988, just as the electricity privatisation process started. At the time, expectations about compliance were conditioned by the pre-existing technological trajectory of the CEGB. This was to anticipate moderate growth for electricity demand and to build as many nuclear power stations as possible. Any remaining capacity gaps would be filled with coal-fired stations. At the time, the way to achieve compliance within the coal-plus-nuclear trajectory was seen to be 12 000 MW of FGD retrofits at existing coal stations. Following privatisation, a paradigm change in the industry's technological trajectory has reduced the perceived need for FGD investment.

Formal compliance is to be achieved under Section 3(5) of the EPA which authorises the DoE to establish *national plans* for emissions reductions. A national plan for sulphur dioxide and nitrous oxide emissions was published in December 1990 just before the flotation of National Power and PowerGen (DoE, 1990b). The plan (see Table 12.3) consists of a set of emission quotas for three different sectors (power stations, refineries and other industry) for each of the three different administrative regions of the UK (England and Wales, Scotland and NI).

The relevant pollution control authority must take account of the plan when setting authorisations for individual large combustion plants. However, National Power and PowerGen have been allocated *company quotas* (Table 12.3) which give them greater flexibility in determining what measures to take at individual plants.

Under the LCPD, the UK is required to draw up a *programme* for the progressive reduction of sulphur dioxide and nitrous oxide emissions from existing large combustion plant. The UK programme incorporates the national plan but, in addition, notes that the electricity industry in England and Wales will retrofit 8000 MW of FGD capacity. While 6000 MW of this capacity is completed or under construction, PowerGen makes no

Table 12.3   UK national plan for sulphur dioxide emissions from existing large combustion plant (ktonnes/year)

	1980	1991	1993	1998	2003
Power Stations	3006	2881	2700	1803	1202
National Power	{2776}	1595	1497	982	660
PowerGen		1085	1019	669	450
Scotland	142	109	104	99	57
N. Ireland	88	92	80	53	35
Refineries	268	100	100	95	90
England/Wales	218	86	86	82	78
Scotland	50	14	14	13	12
Other Industry	621	312	276	230	160
England/Wales	543	273	241	201	140
Scotland	78	39	35	29	20
Total	3895	3293	3076	2128	1452
LCPD Limits			3106	2330	1553

Source: DoE.

mention of the residual 2000 MW, earmarked for Fiddlers Ferry power station, in its recent environmental performance report (PowerGen plc, 1994).

Policy Approaches

Two themes underpinned the UK's position during the negotiation of the 30 per cent club and the LCPD: (a) the need for firm scientific evidence; and (b) the importance of cost considerations in determining whether action is justified (HM Government, 1984). Although the Government's 1990 environment White Paper put a more proactive gloss on these themes by flagging the relevance of precautionary action (DoE, 1990a: para. 1.16), it reiterated the importance of a strong scientific foundation for policy, the best economic information and the use of the best policy instruments. The 'sound science' theme is particularly relevant in international negotiations being held under UNECE/LRTAP auspices. Here, the UK's stress on the need for a scientific basis for policy has been a key factor in countries agreeing to negotiate future emissions reductions on the basis of scientific criteria embodied in the critical loads approach.

WHAT HAS CHANGED IN THE 1990s

The Political Eclipse of Acid Rain

The Prime Minister's September 1988 Royal Society speech raised the political profile of environmental issues in general and climate change in particular. However, this general greening of government policy was followed by a *decline* in political interest in the acid rain issue. By 1993, National Power could discuss its 'emissions and discharges' without using the phrase 'acid rain' (National Power plc, 1993). This decline in interest can be attributed to several factors:

(a)   climate change eclipsed acid rain as the major international environmental issue attracting political attention;
(b)   the issue had apparently been politically tied up by the 1988 agreement on the LCPD; and
(c)   a more general sense of fatigue with the acid rain issue. After several years of media attention, wider interest in acid rain dropped away. In the late 1980s, perceived linkages between hot summers and climate change assumed the kind of potency which images of dying forests had in the early 1980s.

Electricity Privatisation

Britain's positions on the 30 per cent Club and the LCPD were formed within the context of a nationalised electricity supply industry. The process of privatising the industry, coupled with wider changes in energy markets, led to significant changes in the approach to reducing acid emissions. The key changes included the following.

1. First, there was an increase in natural gas availability coupled with a removal of legal impediments to the use of gas in power stations as set by a 1976 European Community Directive. Natural gas has ceased to be regarded as a 'premium' fuel to be reserved for quality markets such as household heating.
2. Improvements in gas turbine technology allowed the development of highly efficient CCGT plant.
3. There was a new commercial freedom for the CEGB successor companies, National Power and PowerGen. They are no longer committed to taking British coal regardless of the price.
4. The plans for new nuclear and coal-fired stations in 1989–90 collapsed following opposition at public inquiries. It became impossible to raise the finance necessary for these long lead-time, capital intensive projects.
5. Finally, there was the introduction of competition into electricity generation. Potential entrants needed technology which would allow them to compete with incumbent generators. The obvious candidate technology, CCGTs, offered zero sulphur dioxide emissions and substantially reduced nitrous oxide emissions. The incumbent generators were forced to abandon traditional technologies and adopt the same cheaper, more flexible options in order to compete with each other and new entrants.

During 1989–90, National Power and PowerGen, operating as 'shadow companies', began to assess CCGTs as an alternative to traditional steam-based generation technologies. It appeared cheaper to invest in new CCGT plant and retire existing coal capacity than to retrofit a coal-fired power station with FGD (Skea, 1990). The industry was committed to at least 6000 MW of FGD following the CEGB's 1986 voluntary decision. The coal industry and its supporters, plus environmental groups, argued hard for the retention of the 12 000 MW FGD programme which had been signalled by the DoE following agreement of the LCPD in 1988 (House of Commons Energy Committee, 1990: 32–5, 41–51). This programme would undermine the environmental advantages of low sulphur imported coal and allow a larger market for British coal.

Following the formal vesting of National Power and PowerGen in March 1990, the formal position became that the two companies would retrofit only 8000 MW of FGD, with other contributions to reduced emissions coming from low sulphur imported coal and CCGT investment. The 8000 MW number was written into both the LCPD programme (DoE, 1990b) and the pathfinder prospectus for the sale of National Power and PowerGen.

RECs, concerned at the potential dominance of National Power and PowerGen in the electricity market, took equity shares in new CCGT projects in order to create an independent generation sector. This type of investment is not environmentally motivated. The architecture of the electricity markets has permitted this strategy to be followed through, even though the resource costs of a new CCGT (omitting externalities) may be higher than the avoided costs of operating an existing coal-fired power station.[4] While National Power and PowerGen were tied into an initial three year contract with British Coal (April 1990–March 1993), new entrants were able to sign up 15-year contracts for gas supplies and effectively lock up a significant proportion of the electricity market for the 1990s and beyond. Planning consent has now been given for 13 800 MW of new CCGT plant, of which 9800 MW is already under construction (National Grid Company plc, 1993). It has not been possible to sign long term contracts for the supply of large quantities of British coal beyond March 1993.

Under the threat of a referral to the Monopolies and Mergers Commission, National Power and PowerGen have accepted the need to lose market share. Inevitably, since they have inherited the stock of polluting power plant, their loss of market share will result in declines in emissions regardless of the degree of clean-up investment undertaken. Figure 12.1 shows throughout the 1990s that there is going to be a very considerable degree of 'overcompliance' with the emission quotas required under the UK national plan.

Many have commented on the irony that the UK spent several years blocking the LCPD on the basis of an erroneous view of the future of the electricity supply industry and, as a consequence, secured relatively lenient emission targets: 'The Government obtained relatively undemanding limits for the UK on the understanding that the UK would achieve the required reductions chiefly through FGD, and that having obtained such limits by that means the UK now proposes to comply with them by cheaper means instead' (House of Commons Energy Committee, 1990: para. 52). Although there is no evidence to suggest that the LCPD negotiations actually took any account of the means by which the UK intended to

Figure 12.1　Projected ESI Sulphur Dioxide Emissions

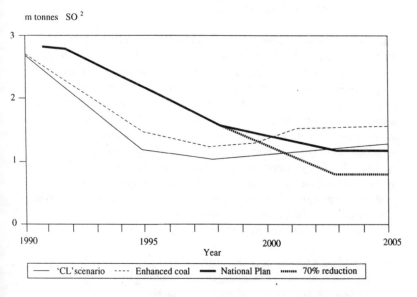

meet emission reduction targets,[5] this observation has led to demands for both a tightening of the UK's emission objectives and a greater degree of investment in FGD (Weir, 1993).

The evidence strongly supports the hypothesis that UK policy on acid rain at the *policy implementation stage* has been determined primarily by concerns about costs and energy market developments, without reference to wider environmental criteria. This has been the case even though policy development took place against the background of a wider 'greening' of government.

Regulatory Institutions

The recent development of acid rain policy has taken place against the background of significant changes in regulatory arrangements for pollution control in the UK. HMIP has a statutory responsibility to implement a system of IPC under the EPA, including the authorisation of power stations and other large combustion installations. Before HMIP issues an authorisation it must be satisfied that the operator's plans are consistent with:

(a)　the national plan for emissions of sulphur dioxide and nitrous oxide;
(b)　European Community Directives and other international law;

(c)   relevant air quality standards; and
(d)   the use of BATNEEC.

The DoE's interpretation of the BATNEEC principle treats BAT and NEEC separately (DoE, undated). According to the DoE, a technology is *available* if it is procurable, anywhere in the world, by an operator. *Best* simply means the most effective in preventing, minimising or rendering harmless emissions. The guidance on NEEC falls into two parts: for new plants, the presumption is that BAT will be used, unless the costs would be 'excessive in relation to the nature of the industry and the environmental protection offered'. For existing processes, the DoE leans heavily on the EC's 1984 Air Framework Directive which requires member states to 'implement policies and strategies, including appropriate measures, for the gradual adaptation of existing plant ... to best available technology' (Commission of the European Communities, 1984: Art. 13). This gradual adaptation should take into account:

(a)   the plant's technical characteristics;
(b)   its rate of utilisation and length of remaining life;
(c)   the nature and volume of emissions; and
(d)   'the desirability of not entailing excessive costs for the plant concerned, having regard in particular to the economic situation of undertakings belonging to the category in question'.

Prior to determining authorisations for individual sites, HMIP develops general guidance notes in order to harmonise the decisions taken by individual inspectors. These 'Integrated Pollution Regulation' (IPR) notes constitute a *de facto* standard for the operation of specific classes of plant. HMIP's IPR note on large combustion plant was developed during 1990 against the contentious background of preparations for the privatisation of the electricity generators (HMIP, 1991). The note was needed to clarify the regulatory demands which the privatised generators would face and went through nine drafts before it was finalised.[6]

The debate over the IPR note centred on (a) the upgrading of existing plant, and (b) nitrous oxide emission limits for new combustion plant. The IPR note suggests that all existing plant should be 'upgraded to achieve the levels ... for new plants whenever the opportunity arises and by 1 April 2001 at the latest', though this objective may be reviewed.

Although it is highly unlikely that new combustion plant falling within the scope of the IPR note will be built for at least a decade, the nitrous oxide limit for new plant was significant because it would define the

BATNEEC towards which existing plant would have to work under the 'gradual adaptation' principle. The IPR note which embodies nitrous oxide standards considerably less stringent than that achievable by selective catalytic reduction (SCR) technology, which is widely used in Germany and Japan, apparently satisfies all the criteria for BAT and would result in a 'relatively small increase in the costs of electricity generation' (DNV Technica, 1992).

In determining authorisations for specific power plants, HMIP has had to deal with: (a) the treatment of orimulsion, a high sulphur crude oil/water emulsion obtained from the Orinoco basin in Venezuela; and (b) upgrading plans. In spite of pressure from the generators, HMIP has defined the conversion of a power station from oil to orimulsion-firing as a substantial change which will require the use of FGD equipment. Following this decision, National Power has withdrawn its application to burn orimulsion at Pembroke. However, PowerGen's remarkably extended trials for burning orimulsion at Ince and Richborough, which began before IPC implementation, are being allowed to continue until 1998 without FGD retrofitting.

Resolving the question of upgrading existing plant to new plant standards by 2001 is likely to be a long-drawn out process. HMIP has asked National Power and PowerGen to submit upgrading plans consistent with the IPR note by April 1994. The results of this regulatory procedure are unlikely to become known for a considerable time as the generators' proposals will not be placed on public pollution registers (ENDS, 1994: 7–8). Strictly interpreted, the IPR note could be said to require the full application of BATNEEC standards (i.e., FGD retrofitting) at all existing coal-fired power stations.[7] However, the generators are likely to argue that the costs of emissions abatement would be excessive in relation to the environmental protection achieved. They can point out that by 2001 (a) existing coal plant will be called on to operate less frequently, and (b) it will have a short residual life.

If HMIP were to insist on upgrading, the electricity generators might simply close their plants.[8] Table 12.4 shows that retrofitting FGD to such stations would push avoidable generation costs up by around 1.5 pence/kWh rather than the 0.55 pence/kWh for long-life base-load plant. Unless gas prices were to rise very substantially, the likely outcome is that such coal plants would be replaced with CCGTs.

International Developments

LRTAP signatories have been negotiating a new sulphur protocol intended to replace the 30 per cent Club, which expired at the end of 1993. However, the deadline was missed and the new protocol was signed only in June 1994.

Table 12.4   Estimated power generation costs (pence/kWh)

	CCGT	Uncontrolled coal	Base load coal with FGD	Part-loaded coal with FGD
Fuel costs	1.50	1.50	1.55	1.55
Capital costs	1.25	–	0.35	1.07
Non-fuel operating	0.25	0.50	0.65	0.93
Total	3.00	2.00	2.55	3.55
OFFER range	2.2–3.0	2.0–2.7		

Notes: Avoided costs are estimated for existing coal plant; OFFER estimates are from Review of Economic Purchasing.

The negotiations have moved away from debate over arbitrary emission reduction targets. Scientific input has come in the form of 'critical loads' for sensitive ecosystems. These are defined as 'the quantity of a substance falling on a given area over a given period which a specified part of the local environment can tolerate without adverse effects occurring'. Least cost methods of reducing acid deposition loads down to the critical levels have been identified. In 1992, countries agreed that it would be impracticable to meet critical loads throughout Europe by early next century. Instead, they have agreed to aim for a 60 per cent 'gap closure' (the gap being that between current sulphur deposition rates and critical loads) for most of Europe by the year 2005 (ENDS, 1993a: 44–6). The UK has played an important role in advancing discussions based on the critical load/gap closure concepts.

As a consequence of the critical loads/gap closure analysis, the UK was asked to commit itself to a 79 per cent reduction in total sulphur dioxide emissions starting from a 1980 baseline. The UK counter-proposed a 70 per cent reduction by 2005 and the full 80 per cent reduction only by the year 2010. Other countries have reluctantly accepted the UK proposal, sensing perhaps that the UK would, yet again, be prepared to stay outside an agreement rather than commit to the earlier compliance date (ENDS, 1993b: 42). Partners negotiating within the LRTAP framework are frustrated that the UK, having promoted science-based policy and the critical loads concept, now appears to be backing away from the consequences in terms of emission reductions (ENDS, 1993a: 44–6). Other European representatives report that they cannot discern any substantial development in the UK's approach to negotiations since the early 1980s and the days of the 30 per cent Club (private communication, 1993).

The UK's prime concern appears to have been the effect of a more ambitious target on coal markets and its adverse effect on the prospective privatisation of coal, together with its effect on the long-term prospects for National Power and PowerGen in which the Government still holds a 40 per cent share. As indicated above, any requirement for additional sulphur dioxide emission reductions beyond the year 2000 is likely to result in further closures of existing coal-fired power stations.

The other problem is that an 80 per cent reduction in sulphur dioxide emissions by 2005 might cause the UK to seek abatement from both large combustion plant *and* other sources as Table 12.5, which extrapolates current sulphur dioxide emission trends, shows.

For power stations and other large combustion plant, emissions reductions would need to go beyond those required under the current national plan. Also, the factors which underpinned reductions in sulphur dioxide emissions from smaller sources in the early 1980s – the collapse in heavy manufacturing, a switch from fuel to natural gas and the growth in the use of sweet, low sulphur North Sea oil – will not be repeated over the next 10–15 years.

THE FUTURE AGENDA

Overview

As in the past, the main pressures to reduce UK acid emissions are likely to be external. There are several identifiable developments within the

Table 12.5   Extrapolation of current trends in UK sulphur dioxide emissions

	1980	1985	1991	2003	2005	2010
Large plant	3867	3043	2973	1452	1289	958
of which power	3007	2627	2534	1202	1061	778
Small plant	1031	681	592	447	427	380
Total	4898	3724	3565	1899	1716	1338
70% reduction					1469	1469
80% reduction					980	980

Note: This is an extrapolation, not a computer model based projection. Power industry sulphur dioxide emission reductions may accelerate beyond 2000.

UNECE and EU frameworks which may have significant implications for UK policy. However, partly through UK influences over negotiations, many of the potential agreements may have only a weak impact.

UNECE

The 1988 Nitrous Oxide Protocol expires in 1994. As with sulphur, the Parties will seek a follow-on protocol to set targets for the year 2000 and beyond. Once again, this will be negotiated using the critical loads approach. Countries are experiencing much more difficulty in securing nitrous oxide stabilisation than they did in reducing their sulphur dioxide emissions by 30 per cent. The shape of any future nitrous oxide protocol is therefore unclear.

European Union

Towards Sustainability

The EC outlined its approach to acid rain in the new Action Programme on the environment, *Towards Sustainability* (Commission of the European Communities, 1992: Section 5.2). As with the UNECE negotiations, considerable reliance is placed on the critical loads concept. The Community aspires to 'the strictest possible measures' and to go beyond a 65 per cent reduction in sulphur dioxide emissions and a 60 per cent reduction in nitrous oxide emissions by 2010. For 2000, specific targets are set for a 35 per cent reduction in sulphur dioxide emissions compared to 1980 levels and for a 30 per cent reduction in nitrous oxide compared to 1990 levels.

If these targets are to be taken seriously, measures other than the LCPD and current vehicle emissions controls will be required. The Action Programme signals the intention of generating proposals for the sulphur content of coal and residual fuel oil by 1995 as one contribution towards achieving the emissions targets.

Integrated Pollution Prevention and Control

A new system of *integrated pollution prevention and control* (IPPC) is one of the key elements of the Union's approach to reducing the environmental impacts of large scale industrial processes. This system would allow impacts on the three main environmental media – air, water and land – to be assessed within a single regulatory framework.

The Commission proposed a draft IPCC Directive in September 1993 (Commission of the European Communities, 1993). This would require existing plant to be brought up to new plant standards by June 2005. However, it specifically excludes existing combustion plant and can therefore have no effect on UK acid emissions. From the point of view of the UK, this arrangement is compatible with its insistence on not reducing emissions by 80 per cent by 2005 under the new sulphur protocol.

Revised LCPD

The 1988 LCPD contained two dynamic elements:

(a) in 1994, the Commission is obliged to review the year 2003 sulphur dioxide emission target, and the year 1998 nitrous oxide emission target for existing plant;
(b) by July 1995, the Commission is obliged to submit new proposals for emission limits for new plant.

The Commission could propose lower emission targets for existing plant on the grounds that UK abatement costs have proved to be much lower than anticipated when the LCPD was first agreed. The UK's emission reduction target might be increased to at least 70 per cent (the level which has been agreed for Belgium, Denmark, Germany, France and the Netherlands).

The Commission's review of emission limits is likely to focus on nitrous oxide. There is a wide divergence in nitrous oxide standards across the Community. By 1995, *new* large combustion plant will fall within the scope of the IPPC Directive, and the Commission is likely to attempt to define SCR as BAT, putting further pressure on the UK.

Small Combustion Plants

The Commission is working on a small combustion plant directive.[9] The lower size limit for such a Directive is undecided, but it could affect both smaller industry and larger service sector establishments. There are two reasons for considering a small plants Directive: first, as controls on larger plants take effect, smaller plants will account for an increasing proportion of acid emissions. Controls will be necessary if the emissions reduction targets indicated in the new environmental Action Programme are to be achieved. Second, tightening controls on large plant could drive dirtier fuels (high sulphur coal and fuel oil) into smaller plant where their use would be undesirable from the point of view of the local environment.

Acid Emissions and the Oil Industry

As the LCPD takes effect, acid emissions from coal/electricity will decline rapidly. However, oil use is projected to rise over the next two decades while the production of low sulphur North Sea crudes will decline. An increasing proportion of acid emissions will therefore originate from oil refining and, more importantly, oil use. The Community has already signalled that product standards for residual fuel oil, which accounts for the bulk of oil-related sulphur dioxide emissions, are likely to be pursued.

All the political and economic problems which have arisen with respect to coal and electricity apply equally to the oil industry. There has been structural overcapacity in the European refining industry since the 1973 oil crisis, and the profitability of the sector is low. Additional costs incurred at refineries, whether to meet changes in the product mix or changes in product specifications, could be a factor in determining exit from the industry.

The costs of complying with a sulphur standard for fuel oil (say 1 per cent) at the Community level would be much higher than those associated with switching to unleaded petrol or complying with the recent Gas Oil Directive.[10] Current UK restrictions on the sulphur content of fuel oil are comparatively lax, and UK refineries could be disadvantaged if Union wide limits on the sulphur content of fuel oil were to be introduced. This could add to UK reluctance to support substantial reductions in acid emissions in the late 1990s and beyond.

CONCLUSION

There has been no substantial change in the flavour of UK acid rain policy over the last decade, though emission reduction *methods* have been strongly influenced by technological and institutional change in the electricity sector.

1.   Policy has been determined as a result of negotiations between a small core group of players comprising the DoE, the DTI and the electricity industry. HMIP has played a peripheral role, while environmental NGOs have remained outside the policy network.

2. UK policy shows no consistency with respect to a particular form of discourse. A science based approach was used in the 1980s and was pushed hard by DoE negotiators within the framework of UNECE negotiations. However, this has been given a lower priority since coming into conflict with other objectives (Murlis, 1993).

3. The factor which most consistently explains policy is the reluctance to incur higher electricity generation costs or perturb energy markets.

4. The 'greening' of government in 1988–89 has had no perceptible impact.

UK policies appear unlikely to change. There was no compelling reason for the UK to accept the new UNECE sulphur protocol on the terms asked for by negotiating partners. The exclusion of existing large combustion plant from the new IPPC arrangements being developed by the EU mean that there will be no 'technology based' imperative to upgrade or close existing plant. HMIP is unlikely to be able to insist that the electricity industry should upgrade or close plant under the IPC system. Any step in this direction will put further pressure on already depleted markets for British coal.

Nevertheless, the revision of the LCPD, small combustion plant controls and further developments within the UNECE framework will continue to put pressure on the UK. Emissions from oil products could add a new dimension to international discussions later in the 1990s.

Notes

1. Jim Skea is British Gas/Economic and Social Research Council Professorial Fellow of Clean Technologies and Industry in the SPRU at the University of Sussex. He is grateful for the support of these two bodies. This chapter was improved by discussions which took place during the Colloquium 'UK Environmental Policy in the 1990s' in the Department of Politics, University of Newcastle, in December 1993. Subsequent correspondence with Nigel Haigh was particularly helpful. Tony Ikwue of SPRU also provided helpful input. However, the author remains solely responsible for the views expressed and for any residual errors.

2. I am grateful to Joe Coffman of the US Environmental Defense Fund for his use of the phrase to describe constituencies which press for particular policy approaches for reasons of power, prestige or ideology as opposed to economic interest.

3. Civil servants have indicated in interviews that the hypothesis that there is any connection between UK acid rain policy and the domestic pollution

control regime is unfounded. Effectively, HMIP's predecessors 'did what they were told'.

4. According to cost estimates made by National Power (ENDS, 1994), generators should in principle be indifferent between FGD and a CCGT, if they are operating under an emissions constraint. However, since both National Power and PowerGen are in a position of 'overcompliance', retaining coal-firing would appear to be cheaper.

5. The DoE believed that the negotiated emission reductions would require a 12 000 MW FGD retrofit programme and briefed journalists accordingly. However, an interview conducted by the author with one of the UK negotiating team within a month of the LCPD agreement, before the later controversy arose, indicated that this understanding was not revealed by the UK during the course of negotiations.

6. A letter from the HMIP Chief Inspector on this subject was included in the electricity prospectus.

7. CCGTs fall within the scope of a separate guidance note on gas turbines.

8. National Power already plans to close 10 of its 18 coal-fired stations by 2000 (ENDS, 1994: 7–8).

9. By Arthur D. Little and the Institut Français de l'Energie.

10. This information comes from interviews carried out as part of a SPRU/)École des Mines project on regulation, innovation and competition in the European refining sector.

References

Boehmer-Christiansen, S. and Skea, J. (1991), *Acid Politics; Environmental and Energy Policies in Britain and Germany* (London: Belhaven Press).

Commission of the European Communities (1984), 'Council Directive on the Combating of Air Pollution from Industrial Plants' (84/360/EEC), *Official Journal of the European Communities*, L 188/20, Brussels, 16 July.

Commission of the European Communities (1992), *Towards Sustainability: A European Community Programme of Policy and Action in relation to the Environment and Sustainable Development*, COM(92) 23 final, Brussels, 17 March.

Commission of the European Communities (1993), *Proposal for a Council Directive on Integrated Pollution Prevention and Control*, COM(93) 423 final, Brussels, 14 September.

DoE (1990a), *This Common Inheritance: Britain's Environmental Strategy* (London: HMSO) Cmnd 1200.

DoE (1990b), *The UK's Programme and National Plan for Reducing Emissions of SO_2 and NO_x from Existing Large Combustion Plants* (London: DoE).

DoE (1993) *Making Markets Work for the Environment* (London: HMSO).

DoE (undated), *Integrated Pollution Control: A Practical Guide* (London: DoE).

DNV Technica Ltd (1992), NO_x *Pollution Control for the UK Power Industry*, Report to the DoE.

ENDS (1993a), 'Government Protects British Coal in Talks on new SO_2 Target', Report No. 222.

ENDS (1993b), 'UK Concession on SO_2 Emission Targets', Report No. 227.

ENDS (1994), 'FGD Arrives – But Future SO_2 Control Policy Remains Uncertain', Report No. 228.

HM Government (1984), *Statement by the UK Delegation at the Munich Air Pollution Conference*, 24–27 June 1984, reproduced in House of Commons Environment Committee, *Acid Rain, Volume II Minutes of Evidence*, Fourth Report, Session 1983–84, HC 446-II (London: HMSO) 302–3.

HMIP (1991), *Chief Inspector's Guidance to Inspectors – Combustion Processes: Large Boilers and Furnaces 50 MW(th) and Over*, IPR 1/1 (London: HMSO).

House of Commons Energy Committee (1990), *The Flue Gas Desulphurisation Programme*, Third Report, Session 1989–90, HC 371 (London: HMSO).

Murlis, J. (1993), 'Critical Loads: The Political Background', Paper presented at the conference 'Acid Rain and its Impact: The Critical Loads.

National Grid Company plc (1993), *1993 Seven Year Statement* (Coventry: National Grid Company).

National Power plc (1993), *National Power Environmental Performance Review 1992* (Swindon: National Power).

PowerGen plc (1994), *Environmental Performance Report* (Solihull: PowerGen plc).

Private communication (1993), Dutch negotiating team.

Skea, J. (1990), *Acid Emissions from Stationary Plant: Reopening the Debate* (London: Friends of the Earth).

Weir, F. (1993), *From 'Dirty Man' to 'Drittsekk': UK Acid Rain Policy* (London: Friends of the Earth).

Wetstone, G. S. and Rosencranz, A. (1983), *Acid Rain in Europe and North America: National Responses to an International Problem* (Washington, DC: Environmental Law Institute).

13 Nuclear Waste Disposal: A Technical Problem in Search of a Political Solution[1]

Andrew Blowers

INTRODUCTION

Radioactive waste management in the UK is in disarray. In addition to the wastes already accumulated from power generation and reprocessing there is shortly to be added a growing volume of decommissioning wastes as one by one the early power stations are shut down and made obsolete or surplus military equipment (submarines, warheads) is dismantled. The opening of THORP will add large volumes of intermediate (ILW) and low level waste (LLW) from both domestic and foreign spent fuel. There is at present no long term disposal route available for the intermediate and high level (HLW) wastes, the bulk of which remain in various states of conditioning in different waste streams. As the wastes have accumulated, the uncertainty about what to do with them has increased. There is a lack of technical consensus over the most appropriate method of managing the wastes. Reprocessing and nuclear wastes have become inextricably entwined as issues of intense political conflict. One by one the options for the management of radioactive wastes have been abandoned in the face of intense political opposition. Only the proposed NIREX deep repository for ILW/LLW at Sellafield remains a practical prospect, but both technical and political issues make its timing and even its existence uncertain. Radioactive waste management is a technical problem in search of a political solution.

There are various dimensions of conflict which, taken together, account for the policy impasse which characterises radioactive waste management.

THE TECHNICAL DIMENSION

Technically there are accepted management routes for ILW and LLW including co-disposal in a deep repository (currently favoured by the UK), subsea-bed disposal as at Forsmark in Sweden or, for LLW only, disposal in shallow repositories as currently practised in France, the USA and at Drigg near Sellafield in the UK. But there is not (and cannot ever be) a proven system of management for the most long-lived wastes. For the longer term there is a broad technical (but not a political) consensus favouring deep geological disposal for HLW whether disposed of as spent fuel or in conditioned form from reprocessing. As yet there are only experimental repositories for these wastes.[2] For long-lived intermediate level wastes (LLILW) there are also experimental facilities.[3]

For the medium term there are a range of options for these wastes. HLW in the form of spent fuel may be left to cool on site either in pools or dry stores or be brought to a single site for monitored retrievable storage (MRS) pending deep geological disposal.[4] Or spent fuel may be reprocessed (as in the UK and France) to separate out the plutonium and uranium from the residual liquid HLW which can then be vitrified and stored. Reprocessing has become an issue of intense political conflict both because it creates large volumes of ILW and LLW which require managing[5] and because it separates out plutonium, so intensifying the problem of nuclear proliferation. The UK Government argues that reprocessing from THORP does 'not add any new waste management issues to those which will already be dealt with' (DoE, 1993: 35), and that reprocessing is a matter for commercial judgement since 'the contractual arrangements between BNFL [British Nuclear Fuels Ltd] and their customers are the best and clearest evidence of the utility of reprocessing and THORP' (Radioactive Substances Act, 1993: 40).

THE ECONOMIC DIMENSION

The nuclear industry remains in the public sector. The putative privatisation of nuclear electricity generation was aborted when financial interests baulked at the possible decommissioning costs. The government currently assumes that financial provisions for decommissioning will be accumulated over a period of up to 135 years at a 2 per cent discount rate. This would minimise the costs to the present generation. But, as critics have argued, there is no certainty either that the economy will generate the necessary growth or that the institutions responsible will survive over such a long time span. By that time there could have been a loss of the expertise and commitment necessary to undertake decommissioning. The utilitarian assumptions built in to discounting tend to undervalue the needs of future generations. Critics claim that a 'sustainable approach gives much more weight to the interests of future generations and requires the avoidance of long-term environmental damage' (MacKerron, Surrey and Thomas, 1994: 11). Decommissioning will remain a major area of contention and a potential barrier to privatisation.

Another area of economic conflict is over the costs and benefits of reprocessing. BNFL places emphasis on the profits from foreign contracts, the investment in jobs both locally and in the UK as a whole, and the prospects of future development (BNFL, 1993). Its adversaries point to the uncertainties in the cost and profit estimates, the future vulnerability of the plant if customers withdraw and the high opportunity costs of investing in THORP, and insist that 'THORP is a plant which has lost its future because in the time it took to build, it lost its original rationale' (Large and Associates, 1993: 9). The government's decision to go ahead with THORP was largely based on the economic case as expressed by BNFL.

THE ENVIRONMENTAL DIMENSION

At the heart of conflicts over radioactive waste are concerns about the environmental impacts. There is little concern about visible impacts such as pollution or resource depletion; indeed, a presumed strength of nuclear energy is that it has less impact on GHG output[6] than fossil fuels. It also avoids the large scale resource depletion and environmental degradation associated with coal, oil and gas based electricity generation.[7] The major concern with nuclear energy is the health risks it poses to present and future generations. It is the impact of routine emissions of radioactivity and the possibility of catastrophic accidental releases that arouses public concern. The general relationship between radioactivity and certain cancers and genetic effects is established, but the specific relationship between particular incidents or nuclear facilities and impacts on individual health is difficult to determine and therefore contentious. Experts can at best only provide cautious conclusions on the basis of evidence. For example, the Black Report could only conclude that a link between leukaemia clusters and the Sellafield plant was 'not one which can be categorically dismissed, nor on the other hand, is it easy to prove' (Black, 1984). Over the years the strong circumstantial but not categorical evidence has been reflected in a series of court cases brought by individuals and environmental groups, often resulting in out of court settlements.

One of the outcomes of this increasing concern has been a progressive tightening of the environmental standards applied to nuclear facilities. For nuclear waste facilities the principles of environmental protection were established over a decade ago (HMSO, 1984). They cover siting, monitoring, waste treatment and control procedures as well as radiological protection objectives. In the UK, radioactive discharges and disposal of solid radioactive wastes are controlled by the relevant inspectorates (HMIP and MAFF in England and Wales, and HMIP in Scotland). Their conclusions must have regard to the dose limits to members of the public from all nuclear facilities set by the International Commision for Radiological Protection (ICRP) and recommended by the UK's National Radiological Protection Board (NRPB). These limits are 1 milli Sievert (mSv) per year from all sources and 0.5 mSv from any one source (0.3mSv from a single new plant and as low as 0.1 mSv for a single radioactive waste repository). In addition the ICRP prescribes that all doses should be as low as reasonably achievable (ALARA), and that 'No practice involving exposures to radiation should be adopted unless it provides sufficient benefit to the exposed individuals or to society to offset the radiation detriment it causes' (ICRP, 1977). In making assessments, the risk of receiving a particular dose must also be estimated in terms of a critical group that is most likely to be exposed.

Clearly the assumptions, calculations and the limits themselves will be matters of contention. Among the potential areas of conflict are: the identification of a critical group; the estimation of the impact of specific radionuclides; the problem of calculating ALARA by giving too much weight to the social and economic costs of achieving marginal reduction in risk; and the lack of estimates of detriment, (i.e., a judgement of the overall severity of impact on an exposed group and its descendants). But the conflicts are not confined to the realm of the experts: risk is a highly politicised issue (Cutter, 1993). The contrast between technical assessments of risks and public perceptions of them is not a matter of expert rationality confronted by innocent emotion. In an area as little understood and contentious as risk analysis, expert judgements are based on values and attitudes; science in these matters is not neutral.

The use of selective evidence to promote a particular viewpoint was amply demonstrated by the main protagonists in the THORP case. BNFL argued that discharge levels had been reduced to such a low level that the most exposed members of the public living around Sellafield received a dose of 150mSv 'similar to that received by taking a return flight to the Far East on a passenger aircraft'. The additional discharge from THORP would be the radiological equivalent of taking one return flight to Tenerife (BNFL, 1993). By contrast, Greenpeace, using calculations of collective dose (the total dose received by a population over a specific period of time) claimed in a rather effective poster campaign that 600 people worldwide would die from the first ten years of operation of THORP.

Environmental risk is an area where the government relies on the advice of expert bodies such as NRPB, COMARE (Committee on the Medical Aspects of Radiation in the Environment) and RWMAC (Radioactive Waste Management Advisory Committee). It is interesting to note the way it dealt with such advice during the consultations over THORP. Two examples may be given. The Chairman of COMARE, in response to the draft authorisations for Sellafield, pointed out that although the limits had been reduced, 'the actual levels of discharge of the majority of radionuclides will be higher than at present' (letter from Professor Bryn Bridges to HMIP, 22 January 1993). Furthermore, he argued that there was no estimation of the detriment caused by THORP which would increase discharges of specific radionuclides. He concluded that 'the possibility cannot be excluded that unidentified pathways or mechanisms involving environmental radiation are implicated. In the light of this, proposals to increase the level of discharge of any specific radionuclide ... should be viewed with some concern'; increased risk to the general public living in Seascale could not be ruled out. Instead of eliciting a detailed response COMARE's concerns were treated in the form of comments and response and some were sent to the NRPB for comment (HMIP, 1993). The issues were taken up by various groups during the second consultation on THORP (DoE, 1993). In addressing these concerns the government's response on the issue of detriment merely noted that international radiation protection was under constant review and that there had been limited progress towards developing an overall index of detriment. They considered 'the current system provides a sound basis for assessing the level of protection' (Radioactive Substances Act, 1993: para 48). On individual dose levels they were content that the discharges would represent acceptable risks (para 53) and on the discharge limits they argued that, though some limits would increase, the overall impact to a critical group 'would be lower than that resulting from discharges at the current limits' (para 65). The point is that the government chose not to address fully the concerns raised by its expert body on the health impacts of nuclear facilities (Friends of the Earth, 1993: 11).

In the second case the government asked RWMAC for an interpretation of collective dose assessments and the validity of presenting such impacts in terms of numbers of fatalities. This was prompted by the Greenpeace claim that 600 people will die as a result of THORP. RWMAC's deliberations (which were subsequently leaked) gave comfort to both sides. On the one hand, the committee concluded that 'the calculation of the statistical number of notional deaths in the whole world population over ten thousand years, as performed by Greenpeace, is reasonably close to the

RWMAC's own calculation' (letter from RWMAC to Minister for the Environment, 5 November 1993). On the other hand, it argued that though collective dose is a useful comparative measure it gives no information about the distribution of doses over place and time, and that the 'use to which Greeenpeace has put collective dose in terms of fatalities ... is misleading'. The RWMAC response was challenged in detail by Greenpeace (letter from Lord Melchett to Minister for the Environment, 13 December 1993), but RWMAC's analysis was used by the government to conclude that 'the average risk posed to individuals of the worldwide population from discharges from Sellafield and THORP are very small compared with the equivalent risk from background radiation', and should 'be balanced against the risks of using other sources of energy supply' (Radioactive Substances Act, 1993: 21–2). It was clear that the government selectively used its advisory body to discredit one of the major opponents of THORP.

THE SOCIAL DIMENSION

The conflict over nuclear risk also has a social dimension. It is claimed that pro-nuclear communities are those familiar with the industry whereas anti-nuclear feelings are easily aroused by proposals for nuclear facilities in greenfield locations where communities have no experience of the nuclear industry (Gittus, 1992). The polarisation of views is certainly exploited by the protagonists. The nuclear industry emphasises the economic benefits (employment, local investment) and community facilities it provides for the local population. This is reflected in the local support for the industry, as revealed in opinion polls. By contrast, environmental groups create and foster anti-nuclear attitudes by exploiting people's anxieties about nuclear power. Put crudely, there appears to be a general, if vague, anxiety among the population at large which contrasts with the high level of support for the industry among its workforce and local communities.

The social components of conflict are, of course, more complex than this superficial polarisation suggests. This is indicated by closer examination of the survey results. Concern varies over time and is likely to be highest just after a nuclear accident or other unfavourable incident (e.g., a report on health effects or an anti-nuclear campaign: (see Rosa and Freudenberg, 1992). For instance, public concern (unprompted) about nuclear waste appears to have drifted down from around 15 per cent in the late 1980s to around 4 per cent in 1992, falling well below concern about many other forms of pollution. Nuclear waste is likely to arouse more hostility than other aspects of nuclear energy because it provides few benefits (little employment) but poses risks which persist far into the future. This explains the apparent paradox that among the opponents of nuclear waste facilities will be found many people who are in favour of nuclear power. Such opposition cannot be easily dismissed as an expression of self-interested NIMBYism, though there will undoubtedly be those content for nuclear waste to be managed anywhere so long as it is not on their doorstep. However, they are willing to make common cause with those who oppose the nuclear industry as a whole. Their alliance focuses on the inadequacies of policy and the premature nature of proposals, and argues the need for a thorough examination of all options. What is often regarded as NIMBYism can also incorporate a case for deliberation over policy, the participation of those affected and a willingness to negotiate (Wolsink, 1993).

Power can be mobilised within individual communities to resist unwanted nuclear facilities. Within the community a united front is forged even though supporters may represent different interests and hold opposing positions and ideologies on other issues and differing views on the need for nuclear power. This alliance not only cuts across conventional class and party interests but invigorates latent patterns of integration based on community and local identification (Blowers and Leroy, 1994). As we shall see, the ability of anti-nuclear interests to mobilise coalitions (both within and between threatened communities) proved a telling factor in the power struggle over nuclear waste in the UK during the 1980s. Such coalitions are ephemeral and disappear once the threat has been removed.

Pro-nuclear communities also exhibit certain characteristics. They tend to be remote and economically marginal in the sense that they are dependent on one industry or suffering high unemployment. Such communities are usually socially homogeneous, lacking social elites and exhibiting a culture of defensiveness. Consequently they lack effective power to influence or resist decisions taken elsewhere that affect their interests. Given these social conditions they become 'pollution havens' (or, in the cases examined here, nuclear oases: see Blowers, Lowry and Solomon, 1991), associated with polluting and risk-creating industries that would be resisted elsewhere. These characteristics, taken together, identify what have been called 'peripheral communities' (Blowers and Leroy, 1994).

The outcome of conflicts over nuclear waste facilities reflects inequalities of power between different communities. It is the powerlessness of peripheral communities to resist risk-creating activities, combined with the power of mobilised coalitions to prevent such activities locating elsewhere, which makes the location of such activities in peripheral communities almost inevitable.

The acceptance of the nuclear industry in peripheral communities cannot be taken as unequivocal support. Attitudes both within and outside nuclear communities are tinged with ambiguity. This 'nuclear culture' has been well portrayed by Loeb when describing the US nuclear complex at Hanford, Washington: 'as security and geography separate complexes like Hanford from those who challenge their mission, ... the same isolation that allows nuclear workers to proceed unquestioning in their tasks also makes it far easier for the rest of us ... to safely distance ourselves from what it is we are supporting' (1986: 230). Recent sociological studies of the French reprocessing complex at Cap de la Hague (Zonabend, 1993) and of Sellafield have uncovered the complexity of feelings and attitudes that co-exist within a nuclear community. The Sellafield study revealed that '"Attitudes", at least towards complex issues like nuclear power in a community, are found to be more open-ended, interdependent, multi-valent and thus less stable and discrete, than normally recognised' (Wynne, Waterton and Grove-White, 1993: 2). The researchers uncovered considerable public ignorance of the nuclear industry even in its heartland, an ambivalence combining a sense of humiliation and stigma, a resentment at lack of adequate infrastructure, with 'a resilient determination to make the best of whatever particular situation exists' (Wynne, Waterton and Grove-White, 1993: 4). Therefore support for the nuclear industry may not be automatic and NIREX in particular may suffer from being seen 'as an outsider, unacceptable elsewhere in the country, and unlikely to generate much permanent local employment' (Wynne, Waterton and Grove-White, 1993: 3).

The social dimension of conflict over nuclear waste intensifies the uncertainty of finding acceptable solutions. The development of cross-cutting coalitions can prevent solutions in new locations and, even in nuclear-friendly communities, acceptability of nuclear waste cannot be guaranteed.

POLITICAL CONFLICT OVER NUCLEAR WASTE

These four dimensions of conflict – technical, economic, environmental and social – have tended, over time, to increase the uncertainty of political outcomes in radioactive waste management. There is no longer a technical consensus over the medium term management of the more dangerous wastes, but a conflict between proponents of reprocessing and dry storage. The economic dimension has begun to have an impact on decision making as the huge costs of decommissioning have been recognised, and in the

future the costs of deep disposal may well be a determining factor in the
timing of a repository. Environmental concerns are the source of public
opposition to the nuclear industry and, in recent years, have focused
increasingly on reprocessing and radioactive waste management. Finally,
opposition to radioactive waste proposals has been intensified by a
growing social mobilisation cutting across class, party and community,
and resentment at being selected as repository sites may well undermine
support in nuclear communities in the future.

The four dimensions changing over time affect decision making over
nuclear waste. The process of decision making has also played a major
part in influencing the outcomes of conflicts. Here three tendencies can be
observed. First, there is the tendency for decision making in the nuclear
field to be secretive, a condition which favours pro-nuclear interests. This
is partly tied up with the military connections of the industry but also
stems from an obsessively secretive and elitist culture of decision making
which is characteristically British. Despite recent regulations to ensure
greater openness in government, a remarkable number of documents relevant to THORP and covered by the Environmental Information
Regulations requiring disclosure appear to have been illegally withheld
(Campaign for Freedom of Information, 1993). Such a culture provides
cover and comfort for industrial interests such as the nuclear industry
which enjoys privileged access to decision makers (Lindblom, 1977;
Blowers, 1984). Within the nuclear communities the industry exercises a
pervasive, but often implicit, power which has been described as non-
decision making by observers such as Crenson (1971). The nuclear industry may also benefit from an inertia in decision making which maintains a
set of policies despite mounting evidence that the policies are no longer
valid. This was the case with THORP, as will be explained later.

Over time this secretive culture of nuclear decision making has been
partly exposed. This is largely a result of the second and contrary tendency, the increase in participation and debate by those opposed to the
nuclear industry that has opened up decision making over policy. It has
occurred in response to pressure from community groups, local government, the media and environmental NGOs. In the early years of nuclear
development opposition was especially focused on the bomb, and movements such as the Campaign for Nuclear Disarmament (CND) attracted a
substantial following. Later, opposition became concerned with the civil
nuclear industry as well, focusing first on nuclear power plants but
increasingly on the back end of the cycle. Initially opposition took the
form of protests, lobbying, use of the media for publicity and support for
individual local campaigns. Latterly the major NGOs, Greenpeace and

Friends of the Earth, have consolidated their impact and developed formidable counter-expertise, employing scientists and lawyers to confront the nuclear industry at every level of decision making.

A third tendency has been the internationalisation of decision making. The International Atomic Energy Agency (IAEA) established in 1956 was orientated to 'accelerate and enlarge the contribution of atomic energy to peace, health and prosperity throughout the world'. It has paid some attention to the problem of nuclear waste, but its role is advisory. Similarly the ICRP sets advisory international radiological protection limits to which national governments are expected to conform. The nuclear industry is still substantially under the control of national governments which are free to develop their own standards, locate nuclear facilities where they please and to determine the methods of management. It was noticeable that the UNCED Conference at Rio paid little attention to radioactive waste, and the chapter in *Agenda 21* is timid and placatory towards national interests (UN, 1992). Despite this it seems likely that the international dimension will have a bigger impact on national decision making. International concerns are likely to focus on the trade in radioactive materials, plutonium, uranium and wastes which arise from commercial reprocessing in the UK and France.

These characteristics of UK nuclear decision making – its secrecy, the growing influence of environmental groups and the emerging impact of the international context – also contribute to the tendency towards conflict and uncertainty in policy making. Radioactive waste management policy can be explained in terms of the interaction of the technical, economic, environmental and social dimensions which create the political environment for decision making. The following brief historical overview of the UK's policy shows how the political environment of decision making changed over time in response to the changing dimensions of conflict. It leads on to the analysis of the major contemporary conflicts over reprocessing and the proposed NIREX deep repository at Sellafield.

A HISTORICAL OVERVIEW OF UK RADIOACTIVE WASTE POLICY

Early Phase

During the early years (1940s to 1960s) the emphasis was on military uses and on nuclear energy generation. Radioactive wastes were left on site, dispersed or dumped at sea. A clean-up problem has been left, especially from badly managed wastes on nuclear reservations in the USA, in Russia and (to a lesser extent) at Sellafield. During these years radioactive waste was regarded as a purely technical issue and was of low priority.

The 1970s

As wastes accumulated and environmental concerns increased, attention was focused more on the back end of the nuclear cycle. The issue became politicised but mainly involved experts, politicians and some NGOs. The possibility of the UK becoming a dumping ground for foreign wastes was an issue at the Windscale Inquiry into THORP (1977) and radioactive waste was the subject of the landmark 6th Report of the Royal Commission on Environmental Pollution which urged that 'there should be no commitment to a large programme of nuclear fission power until it has been demonstrated beyond reasonable doubt that a method exists to ensure the safe containment of long-lived, highly radioactive waste for the indefinite future' (Royal Commission on Environment Pollution, 1976: para. 27).

The 1980s

During the 1980s political conflict over radioactive waste broadened and deepened. Following the Royal Commission report there was a technical consensus over methods of management, but technical options met with resistance from NGOs, local government and communities mobilising broad coalitions of support. The government and the nuclear industry fought a series of defensive actions, surrendering options one by one. First, the search for sites for exploratory drilling for potential HLW disposal was abandoned in the face of local community opposition. The government declared that HLW should be vitrified and left for about 50 years. As a result the UK is almost alone in not pursuing a site selection strategy for HLW. The next target of the opposition was sea dumping, which the UK had organised in the north east Atlantic for many years. Greenpeace engaged in spectacular protests, lobbied the London Dumping Convention to achieve a moratorium on dumping, and influenced the trade unions to withdraw from the 1983 annual sea dump. The UK has observed the moratorium. In 1986 the DoE's report on best practicable enviornmental options for radioactive waste management claimed that 'Economically and radiologically sea disposal could be the preferred option for about 15 per cent of ILW' (DoE, 1986). Sea dumping demonstrated the significance of NGOs and the importance of the international context of radioactive waste policy.

Efforts to secure land-based sites for the disposal of ILW and LLW also encountered strong opposition. NIREX announced sites for disposal without any prior warning or consultation. Proposals for a deep repository in an abandoned anhydrite mine at Billingham were withdrawn in 1985 after intense local action persuaded Imperial Chemical Industries, the owners of the mine, to withdraw their cooperation from NIREX. A proposed shallow disposal site at Elstow (Bedfordshire) was opposed by the local councils using scientific counter-expertise and a local action group mobilising across the community. The government conceded a comparative site evaluation and added three more potential sites (in Essex, Lincolnshire and Humberside) to Elstow. This prompted a united front among three county councils and the mobilisation of cross-cutting alliances among the threatened communities. The power of these coalitions was sufficient to persuade the government to withdraw to avoid 'intermediate level embarrassment in advance of an imminent general election' (*The Times*, 2 May 1987). There were no options left and the government and NIREX had to begin the search for solutions again, recognising that public acceptability was now a pre-requisite.

Co-disposal of ILW/LLW in a deep repository was put forward as the technical solution, and also because putting both categories of waste in one instead of two repositories would save overall costs. In a deliberate effort at open and consultative decision making, a wide range of siting possibilities were presented in a report, *The Way Forward* (NIREX, 1987). Responses to the consultation revealed little consensus or support for specific sites, apart from some support in the two nuclear oases of Sellafield and Dounreay. Although NIREX considered 12 sites, only these two were put forward since, the Secretary of State argued, 'it would be best first to explore those sites where there is some measure of local support for civil nuclear facilities' (*Hansard*, March 1989).

Opposition to the proposals was especially effective in Caithness but the decisive factor in the selection of Sellafield was cost. With 60 per cent of the waste arising at Sellafield, a repository in the vicinity would avoid expensive transportation to the more remote site at Dounreay. Thus a political solution had been found within a range of technical possibilities. The Sellafield site had to satisfy the technical criteria and the long process of exploratory and research work began. It was a political option in search of a technical solution.

Throughout the 1970s and 1980s various shifts in the political environment surrounding radioactive waste may be observed. The technical consensus was broken by opposition but was eventually restored with a narrower remit for deep disposal. The economics of radioactive waste began to influence policy, as the preference for Sellafield shows. The environmental arguments were instrumental in mobilising opposition to sea dumping and on-land repositories at greenfield sites. In social terms this mobilisation had incorporated a variety of interests and created coalitions cutting across class, ideologies and communities. In terms of decision making the nuclear industry possessed the advantage of privileged access through the hidden processes of government and commanded sympathy among a majority of politicians and trade unions. But nuclear waste (including decommissioning) was the industry's weak link, and was exploited by the growing influence of NGOs. The international dimension had emerged over the issue of sea dumping. Each of these elements – secrecy, the influence of NGOs and the international dimension – were major aspects of the decision over the THORP reprocessing plant which had subsumed the issue of radioactive waste by the end of the 1980s.

The 1990s

Reprocessing and Radioactive Waste

Reprocessing has been carried out at the Sellafield site for both military and civil purposes. Reprocessing began in 1952 and has been operating for the Magnox programme of power stations for nearly 30 years. It is regarded as technically the best option for managing this particular fuel. Some 35 000 tonnes of uranium fuel (tU) of Magnox fuel from mainly UK (but also foreign) sources has already been reprocessed, and it is estimated that this will rise to around 50 000 tU by the time the programme is completed in the early years of the next century. The large volumes of wastes from the Magnox stations will continue to dominate the total waste burden for a long while to come, and provision must be made for their safe intermediate and long-term management. This long-established programme has been regarded as inevitable and therefore relatively free from political controversy.

By contrast, THORP has become deeply controversial. Conceived as the next phase in reprocessing to deal with Advanced Gas-cooled Reactor (AGR) and Pressurised Water Reactor oxide fuels it was granted planning permission in 1978 after the long-running Windscale Inquiry. In the fifteen years it took to complete the plant there occurred a sea-change in the strategic, technical, economic and political circumstances surrounding the plant. The strategic case for separating out plutonium for the nuclear arsenal fell away with the strategic arms limitation agreements and the ending of the Cold War. Security based on mutually assured destruction was replaced by the fears of nuclear proliferation in a more fragmented and less stable world. The world plutonium surplus of 300 tU in the civil sector and 1000 tU from military programmes became a threat. THORP would add 3 tU a year to the around 70 tU already stockpiled at Sellafield. The UK Government is satisfied that the plutonium is fully safeguarded to prevent theft, sabotage and diversion.

The potential use of recycled plutonium and uranium as fuel had largely disappeared by the early 1990s. The fast breeder reactor programmes which were expected to use plutonium as fuel have been either abandoned or suspended in the UK, France, Germany and the USA. BNFL claims that mixed oxide fuel (MOX) for use in existing and new reactors has great potential (BNFL, 1993), and it intends to commission a plant for its production. Critics argue that MOX is expensive relative to standard fuel, creates more plutonium and produces difficult waste streams (Friends of the Earth, 1992).

The case for THORP has come to rely on its presumed economic benefits to the UK and the West Cumbrian economy and on its role in radioactive waste management. It is argued that reprocessing reduces the volumes of the HLW (by about half); and, by recycling existing fuel, it avoids the environmental degradation, risks and waste that result from mining fresh uranium. By concentrating waste streams at one or two sites, reprocessing reduces the costs and risks of storage at scattered sites. Since the Magnox programme already operates and produces the biggest waste volumes THORP would add no new problems, at least in the early years. If THORP did not exist then new arrangements would have to be made, 'including research and development, for the direct disposal of fuel in the UK or in seeking some other long term solution' (Government statement, July 1993). Indeed the government, following BNFL's claims, argued that THORP was necessary to deal with AGR fuel which had been prepared for reprocessing, some of which was reported to be corroded.

All these arguments have been challenged by opponents, notably Greenpeace and Friends of the Earth. Although reprocessing does not increase radioactivity, it does create very large volumes of ILW and LLW. Indeed, if eventual decommissioning wastes accruing from the reprocessing plant and allied facilities are included, then it is estimated that the volumes amount to 189 times the original fuel assembly (Large and Associates, 1992). Eventually THORP will be contributing 34 per cent of the operational waste arising in the UK and will account for 29 per cent of all decommissioning wastes (Large and Associates, 1993).

Opponents argue that the alternative of dry storage is a cheaper and safer option. This view derives support from Scottish Nuclear, which has opted for dry storage at its Torness power station. Their Chairman in 1991 stated that, 'We have concluded that, based on present costs, reprocessing is uneconomic and unnecessary at the present time.' Officials at the Torness Inquiry accepted that 'the reprocessing route did not appear to offer any immediate and significant advantages from a waste management point of view' (Hickman, 1993). In addition, the government's advisory body, RWMAC, noting the increasing volumes of waste released by reprocessing, concluded that, 'There are ... no significant waste management reasons for or against early reprocessing of spent oxide fuel' (RWMAC, 1990: 44). RWMAC found dry storage to be 'consistent with sound radioactive waste management practice'.

It was the economic and social case for THORP that provided the main justification for the go-ahead. BNFL emphasised the profit over the first ten years of operation, following £2.5 billion investment funded by £9billion worth of orders (half from overseas). The overall benefit to BNFL was calculated at £1.8 billion over ten years (at 1993 prices with an 8 per cent discount rate), and to the UK at £900m. The plant would provide 5450 full-time jobs, over 3000 in West Cumbria. The local economy would benefit from £60m spent annually. The government backed Sellafield on economic grounds in a resolution on 28 June 1993 signed by the Prime Minister, which effectively signalled the go-ahead for THORP.

These economic benefits were questioned by opponents. They pointed out that the analysis was based on a secret report to BNFL and that the opportunity costs of opening THORP had never been properly considered. They suggested that the decommissioning costs had been underestimated; that delays in the return of wastes would put up storage costs; and that any delays to the NIREX repository would also increase costs. There would be downward pressure on prices as alternatives became more attractive or as customers withdrew altogether. The economic risks would increase as time went on, leading to the possibility of premature closure of the plant (Berkhout, 1993, 22).

Nevertheless, start-up was approved in December 1993. The Secretary of State for the Environment, John Gummer, announced that he 'had come to a judgement that there is a sufficient balance of advantage in favour of the operation of THORP and we are satisfied that the activities giving rise to the discharges permitted by the authorities are justified' (statement on new authorisations for the Sellafield site, December 1993). This decision may be explained in terms of the dimensions identified earlier. At the *technical* level there was a breakdown of consensus favouring reprocessing, but the existence of the Magnox reprocessing and the orders for the THORP plant provided a continuing justification. In *environmental* terms the specific decision was to authorise the radioactive discharges at the site. The proposed authorisations gave rise to substantial objections during the initial consultation during November 1992 and January 1993 when 80 000 replies of all kinds were received (77 per cent against the proposals to revise the authorisations and the operation of THORP: HMIP, 1993). During the further period of consultation (August to October 1993) over 40 000 responses were received, and 63 per cent opposed the operation of THORP (DoE, 1993). But the government upheld the inspectorates' original conclusion that the authorisations 'would effectively protect human health, the safety of the food chain and the environment generally' (HMIP, 1993: 7). The environmental issue was sufficient to cause a major controversy and delayed the opening of the plant while intensive consultations were undertaken.

The *economic* and *social* issues appear to have been decisive. The existence of foreign contracts which, together with domestic orders, would ensure the plant's first ten years and the possibility of future contracts was persuasive. THORP also represented a continuing commitment to the nuclear industry, and especially to West Cumbria. The loss of jobs and investment in the area would have severe local and regional ramifications and clearly weighed heavily in the decision.

The decision also reflected inertia in the political environment The decision making process provided an advantage to BNFL with its close contacts within Whitehall, a position it exploited thoroughly. Although the environmental NGOs had created publicity and challenged the policy using scientific expertise and legal processes, they were, ultimately, unable to deliver the votes that mattered. The international influence on the decision was weak and provided support for both camps. The foreign utilities backed the plant while it was opposed by environmentalists in other countries on grounds of the environmental risks and dangers of proliferation. But the plant had been built and was ready to open. Throughout its period of construction although the justification weakened, any decision to review or abort the plant was avoided. By the time it was finished it was too late to withdraw. It was a decision that reflected inertia. Despite the widespread doubt and vigorous opposition from NGOs the plant still commanded a healthy margin of support in Parliament (the June 1993 motion was carried with a majority of 114). It may have been the wrong decision, as many believed, but it was, in the circumstances of the time, the only really feasible decision politically. It was easier to manage the sullen acceptance of the decision to go-ahead than to face the turmoil that would have followed in West Cumbria if the plant had been halted. The plant's opening raised a set of new questions for the future of radioactive waste management policy.

Substitution and the Problem of Foreign Wastes

Sellafield is both the centre of the UK's waste management problem and the site of its potential solution. It has stockpiles of plutonium and uranium, spent fuel in ponds awaiting reprocessing, vitrified high level wastes in store pending eventual deep disposal, encapsulated ILW, and a whole range of waste streams stored in various conditions. The bulky low level wastes are disposed of in the nearby Drigg repository which has been upgraded to take conditioned, compressed and packaged wastes. Excluding Drigg the total waste volumes stockpiled at Sellafield in 1991 were as follows: $1700m^3$ HLW; 53 $100m^3$ ILW and $7600m^3$ LLW (a total of 62 $457m^3$). By around the year 2030 the total volume of radioactive wastes in Britain, based on currently envisaged production and including decommissioning, is forecast to rise to around 1 640 $00m^3$, or about fifty times the present level (Large and Associates, 1993: 2/3).

With THORP opening, an increasing proportion of the wastes arising at Sellafield will be of foreign origin. At present there is only a small volume of foreign originating wastes coming from the Magnox plant. Large and Associates (1992) estimate that as much as two-thirds (about 11 000tU) of the spent fuel reprocessed at THORP will be of foreign origin and that this will contribute around a fifth of the total stockpile of operational and decommissioning wastes. But, since 1976, all foreign contracts have options to return all wastes to sender, and the government has stated that 'the options should be exercised and that the wastes should be returned' (Mr Goodlad reply to Mr Mitchell, 2 May 1986, *Hansard,* col. 502–3). This suggests that all but the wastes from the 1500tU of pre-1976 foreign spent fuel will be returned. This is not, in fact, the case. In the first place it is highly unlikely that all the bulky LLW will be returned. These do not constitute a significant hazard, can be accommodated at Drigg and would involve high costs of transport if they were to be returned. Secondly, not all the wastes arising from foreign contracts are included in the specifications. BNFL suggest about 8 per cent are not covered while Large and Associates (1993) argue that, if decommissioning of THORP and operational wastes (from refits and overhauls) attributable to foreign contracts are included, then as little as 11 per cent of the total volume (though a much higher proportion of the radioactivity) would be repatriated.

A third possibility is that arrangements may be made to substitute HLW for an equivalent amount in radioactivity of ILW/LLW. Thus small volumes of HLW would return while large volumes of ILW/LLW would remain in the UK. The advantage of this would be reduced transport costs, smaller volumes in transit and it would reflect the radioactive waste management routes being prepared for HLW in other countries and the intended ILW facilities being prepared by NIREX in the UK; but it would mean increased volumes of wastes to be managed (Large suggests that under substitution only 0.29 per cent of the total volume of overseas fuel reprocessing wastes will be returned).[8] Although the proposal is neutral in terms of radioactivity it would mean the UK would become the final resting place for the bulk of the wastes arising from foreign contracts.

RWMAC was asked to comment on the implications of substitution. Their reply raised a number of concerns. Among them were that substitution would depend on the development of a NIREX repository that could take all these wastes. They pointed out that substitution would mean that plutonium depleted HLW was being returned, whereas plutonium contaminated ILW would be retained in this country. This 'is an issue of public concern and has the potential to create heated debate over the UK being labelled the "European dumping ground for radioactive waste"' (RWMAC, 1992a). They concluded that the earnings from reprocessing would have 'to be balanced against the cost and acceptability of the increasing amounts of overseas plutonium remaining in the United Kingdom either in store or in the foreign plutonium-contaminated ILW retained'. The advice was not initially published though its implications were discussed in the Press and were the subject of acrimonious and unpublished correspondence involving BNFL and RWMAC.

The government, following BNFL's lead, concluded that, 'Given that substitution of high level waste for low and intermediate level waste is not necessary for THORP to operate, and would only be implemented on terms which are not considered detrimental, it does not need to be considered further here' (Government statement, 1993: para 36). But substitution does raise issues for radioactive waste management. As the Chairman of RWMAC observed, 'One of the fundamental tenets of good waste management practice is that waste should not be generated until a disposal route has been identified.' This much had previously been conceded by BNFL who had stated that, in order to achieve public acceptability, 'It would be helpful to have a firm waste return programme before THORP is operational' (BNFL, 1992: 20, para 14.7). If substitution were to go ahead it would confirm Britain's role as a permanent repository for foreign waste. Public reaction was a factor that had yet to be considered and, as a recent study has shown, there could be considerable resistance and resentment at the idea of the UK becoming an international dumping ground for radioactive waste (RWMAC 1992a; and Wynne, Waterton and Grove-White, 1993).

Even if substitution does not proceed there is no certainty that the foreign wastes will be sent back. The first shipments of foreign wastes back to their country of origin are timed for 1996 and so the

implementation of the return-to-sender policy has yet to be tested
Receiving countries must have appropriate facilities in place and secure
transportation routes. This may cause delay in the programme for the
return of wastes. Although foreign governments may be anxious to abide
by contractual arrangements (whether or not substitution is applied), there
may be public reaction and resistance to the transportation of nuclear
waste through ports, on the high seas and along inland transportation corridors. The controversy surrounding the shipment of 1700kg of plutonium
in the *Akatsuki Maru* from Cherbourg to Tokai in Japan during December
1992 and January 1993 demonstrated the strength of reaction that could be
generated against the long distance transfrontier movement of nuclear
materials. The ability of anti-nuclear groups to mobilise opposition on an
international basis is a factor that must increasingly affect the feasibility of
policy.

Given the delays in repatriation of wastes likely to occur and the possibility that it may be prevented altogether in some cases, the UK must find
ways of managing an indeterminate volume of foreign wastes for an
undefined period. This is in addition to the spent fuel, plutonium, and the
multitude of radioactive waste streams already accumulated at Sellafield. As
Large and Associates have observed, 'Introducing the radioactive outpour-
ings of THORP to an already over-stretched radioactive waste management
system could be ... the final straw breaking the camel's back' (1993: 15).
The whole radioactive waste management problem now relies on the successful opening of the NIREX repository early in the next century. The
problems surrounding this project add further to the uncertainty that is characteristic of the UK's radioactive waste management policy.

The NIREX Repository: An Uncertain Prospect

The NIREX repository must satisfy the technical, environmental, economic and social criteria discussed earlier if it is to succeed politically. At
the technical level it is necessary to construct a substantial underground
vault at around 650m deep, accessed by tunnels from the Sellafield site
(NIREX, 1991). Although some doubt has been expressed about the engineering difficulties likely to be experienced in constructing the repository (RWMAC, 1992b), most critical attention has focused on the
conceptual design, the multi-barrier approach using both engineered and
natural defences to provide physical and chemical barriers against the
migration of radionuclides. The main concerns are the durability of the
packaging and the concrete barriers used to contain the wastes; the feasibility of retrieving the wastes should serious leakage occur; the security of

the chemical barrier provided by the backfilling material; the problem of gas pathways; and the prospect of natural or human disturbance of the repository (NIREX, 1993; RWMAC, 1993).

While these technical problems of the repository itself must be satisfactorily resolved, the environmental safety and security of the proposal will depend on the integrity of the whole system including the surrounding environment. Since the NRPB requires the provision of the same level of protection to future generations as is afforded to people today (NRPB, 1992: 41) some of the long-lived radionuclides placed in the repository must be prevented from reaching the accessible environment for very long time scales, in some cases over 100 000 years. The migration of radionuclides will depend on the speed of breakdown of the wastes, their solubility and absorption characteristics but, most crucially, on the speed of flow of the groundwater passing through the repository. NIREX have undertaken a massive geological and hydrogeological research programme including geophysical surveys, the drilling of ten deep boreholes and are planning to excavate a rock characterisation facility (RCF) in order to ascertain the nature of the rock and water flows in the area of the repository. A series of reports has been published which shows that the flows are complex, determined by the faulting system and in some areas flowing upwards under pressure.

NIREX still retains a confident posture arguing that the site 'continues to hold good promise', but is also notably cautious insisting on the need for more research 'before the level of confidence in potential site performance is sufficiently high to underpin a decision on a planning aplication for repository development' (NIREX, 1993: 40). Critics are less sanguine. RWMAC pronounces it 'an open question as to whether the observed variability in the hydrogeological conditions at Sellafield will provide unequivocal evidence that the stringent hydrogeological conditions required for a deep radioactive waste repository can be met at this site' (RWMAC, 1993: 17). Cumbria County Council considers groundwater may be forced through the rocks at a higher rate than post-closure safety would allow (Cumbria County Council, 1992).

At the very least the need for further investigation to remove some of the uncertainties will delay the repository. Already the time-scale has shifted. Initially, NIREX intended to put in a planning application in 1992 for a repository opening in 1996/7. The timing has slipped and, with the introduction of the RCF as a further stage in the investigations, NIREX suggested 2006/7 as the target date for opening. The RWMAC have argued that there is too little time built in either for adequate monitoring of the RCF or for the inevitable delays in the planning process. They

estimated 2010 as the earliest opening date, and this has now been conceded by NIREX. A substantial delay in the repository (to 2015) could
mean that the Drigg site will be unable to cope with the continuing high
volume of LLW (including possible substituted wastes) unless supercom-
paction is operating, the site is extended or decomissioning is delayed.
But, the biggest problem would be managing the accumulating ILW,
including foreign origin wastes, which will require the construction of
stores. This problem will intensify if substitution is introduced, thus
adding to the volume of ILW requiring management. Insofar as the
NIREX repository cannot accommodate LLILW (because of groundwater
return times), these wastes (including plutonium contaminated wastes)
may have to be stored for much longer periods until an appropriate method
of management has been secured.

There is, of course, no certainty that the NIREX repository will ever be
built. After further investigation the site may prove too uncertain for a
safety case to be made or it may fail at the inquiry stage. Indeed, it may
never be constructed for economic reasons. The scale of the project is
comparable to, say, the Channel Tunnel or a major airport expansion but,
unlike these, it is not one that is likely to attract private investment. It will
be vulnerable to public expenditure restrictions resulting in delay, a more
modest project (e.g., for ILW only) or even outright cancellation. THORP
escaped the Treasury's scrutiny; NIREX may not be so lucky.

It is also possible that the NIREX repository will be abandoned as a
result of social factors. Unlike THORP, NIREX offers few jobs but introduces a potential risk to the environment and health that extends
indefinitely down the generations. Consequently, the repository is not even
noticeably welcome in West Cumbria (Wynne, Waterton and Grove-
White, 1993), though it was specifically chosen as the politically most
promising option. Sellafield was chosen largely because the project was
unacceptable anywhere else. The ability of communities elsewhere to act
in their own defence provided the political context for Sellafield to be
selected as a potential site. Once that decision was taken, interest in the
issue waned and community action groups dispersed. The social and political dynamics that secured the selection of Sellafield will not remain to
ensure the repository is built there many years later.

With the THORP issue now settled a planning application for a repository at Sellafield will provide the next target for the opponents of the
nuclear industry. It may prove a much softer target than THORP. Lack of
local support, the threat to future generations and the dumping of foreign
wastes in the repository are all factors that will play into the hands of an
opposition whose ultimate objective is the demise of the nuclear industry.

If Sellafield, too, eventually proves a no-go option the industry will liter-ally have nowhere else to go in its search for a permanent solution to the radioactive waste problem.

CONCLUSION

The UK's radioactive waste policy is riddled with uncertainties. The waste burden continues to pile up but the types of waste, their location, their volumes and their methods of management will depend on a whole range of unknowns. These include the rate of commissioning of new power sta-tions; the timing and mode of decommissioning; the impact of dry storage; the introduction of methods of compaction and other forms of waste treat-ment; the domestic market for reprocessing spent fuel; the economic and political conditions affecting foreign reprocessing contracts; and the avail-ability and location of potential storage and disposal sites. We simply do not know how much waste, and of what type, will have to be managed, and for how long. Above all, we cannot be certain that a permanent solu-tion to the problem of radioactive waste management will be found.

Neglect of the problem of radioactive waste has long been characteristic of an industry which has been primarily concerned with the challenge of making bombs and building power stations. Now that the focus has neces-sarily shifted to the rear end of the nuclear cycle, conflicts over policy have emerged to create conditions of continuing uncertainty. At the *technical* level there is no longer a consensus on reprocessing, and dry storage is now perceived as an acceptable alternative form of waste management. While deep disposal remains the technically agreed method of long term manage-ment, there are emerging doubts over the hydrogeological suitability of the Sellafield site. The *economic* dimension (in terms of foreign orders and local impacts) was critical in the decision to proceed with THORP but economics have, so far, played a relatively unimportant role in the debate over the NIREX repository. Resources have been provided for the site selection process and for the increasingly expensive exploratory work at Sellafield. But, if the site is eventually found to be technically and environmentally acceptable, the costs of actually constructing it may prove to be a major stumbling block to implementation. *Environmental* factors are also likely to contribute to delay and indecision. Environmental concerns motivated the end of sea dumping and the abandonment of HLW drilling, and focused the concern of the communities who resisted the NIREX proposals during the 1980s. Concern over the environmental impacts of THORP were sufficient to cause a long and unanticipated consultation process. Environmental issues are also likely to play a major part in the decision over whether to proceed with the NIREX repository at Sellafield.

Social factors have played a major part in determing the course and outcomes of UK radioactive waste management policy. During the 1980s the spontaneous (if ephemeral) development of alliances within and between communities was sufficient to defeat a series of proposals, including HLW drilling, deep disposal at Billingham, shallow burial at four sites in eastern England and the possibility of a deep repository in Caithness. These alliances were founded on traditional horizontal patterns of social integration based on community, neighbourhood, local or regional identification that had been supposedly annihilated by the process of modernisation with its vertical but divisive patterns of class solidarity (Blowers and Leroy, 1994). The corollary of successful resistance was the inevitable selection of Sellafield, a peripheral community, internally fragmented and thus unwilling and unable to mobilise a united front against the further development of the nuclear industry. But the local support for THORP, based as it was on economic dependence, will not necessarily hold for NIREX, a project which brings no lasting benefits while imposing potential long term hazards on the locality.

Should the NIREX project fail for whatever reasons – lack of safety, public expenditure restrictions, local opposition – it is difficult to envisage politically acceptable alternative mainland sites. Opposition could again be mobilised against any greenfield site, but the decision will also be affected by two new *political* dimensions which are already emerging.

First, there is the tendency for conflict over nuclear issues to be increasingly engaged by opposed unaccountable interests. During the 1980s the communities opposing NIREX tended to work within the system of representative democracy: however, in the case of sea dumping and later of THORP, Parliament played a relatively passive role. THORP became a battle between two protagonists, BNFL and Greenpeace. Each side claimed to represent certain interests: in BNFL's case, the interests of economic wealth and employment, while Greenpeace assumed the role of guardian of environmental values. Neither was 'representative' nor accountable but both could claim to be acting on behalf of specific communities of interest. To an extent decison making over nuclear issues is becoming captured by opposing interests, each attempting to mobilise the expertise and support necessary to defeat the other. Environmental politics, of which radioactive waste is a part, is becoming the domain of rival, powerful lobbies and experts rather than an arena of popular participation.

Second, this lack of participation is reinforced by the nature of decision making in the UK. The obsessive secrecy provides privileged access to powerful business interests which, for example, BNFL were able fully to exploit over THORP. Government seeks advice from its nominated advisory bodies (such as RWMAC and the Advisory Committee on the Safety of Nuclear Installations in the case of the nuclear industry), which tend to work in secret despite the presumption of openness required by the Environmental Information Regulations of 1992. This culture of secrecy has spawned a complementary tendency of leaking confidential papers to environmental groups or the media from within a government department or advisory bodies. By acquiring confidential information and by deploying counter-expertise, environmental groups have begun to penetrate the decision making process. One of the noticeable features of the THORP conflict was the way in which Greenpeace gained legal 'standing' and thus became regarded as a body legitimately representing the interests of its supporters. As a result conflict between environmentalist and nuclear interests is increasingly becoming conducted within government, often between government departments. Paradoxically the success of Greenpeace in gaining access to the decision making process may be to institutionalise the conflict and discourage wider participation in open public debate.

Third, the conflict has also taken on an international dimension. Hitherto, sea dumping apart, international controls over radioactive wastes have been relatively weak and, on the whole, individual countries are left to devise their own policies. As the trade in radioactive wastes resulting from reprocessing facilities at La Hague and Sellafield increases, however, so the conflict will become increasingly internationalised. Both the nuclear industry and its opponents are able to mobilise support that transcends national boundaries. For example, during the THORP debate environmental networks opposing the plant emerged linking groups in Japan, Germany, the USA and the UK. They were countered by international linkages between nuclear interests. For example, ten Japanese power companies took out advertisements urging the UK Government to go ahead with the plant. In the future, decisions over radioactive waste will be increasingly constrained by international pressures.

Since the late 1970s the UK's radioactive waste policy has been shaped by political pressures reflecting a combination of economic, environmental, technical and social factors. These have led to the deferral of any long-term solution for HLW and to the selection of Sellafield as the only option for dealing with ILW and LLW, including a substantial burden of wastes of foreign origin. The future of the NIREX repository will be decided not only by its domestic political context but by an international context in which the major contestants are able to mobilise and deploy resources, expertise and political power within and between the nuclear nation states. In that context the outcome of policy making will continue to be unpredictable and uncertain.

Notes

1. This was originally produced as part of a research project on Setting Environmental Agendas in the ESRC's Global Environmental Change Programme.
2. For example, the international experimental facilities at Lac du Bonnet in Manitoba, Canada, and at the Stripa mine in Sweden.
3. As at Carlsbad, New Mexico, USA, which is intended to become a permanent repository if approval is granted.
4. The Centrala Lagret fër Avant Karnbrasle (Swedish Interim Spent Fuel Storage Facility) store in Sweden is an example of an MRS.
5. Estimates vary between 60 and 189 times the volume of the original spent fuel. See Large and associates (1993), and DoE (1993: 39).
6. Although the use of fossil fuels in constructing nuclear energy facilities is usually overlooked in such promotion.
7. There is, nevertheless, considerable degradation associated with uranium mining.
8. Large calculates that BNFL only includes 11 per cent of the foreign originating wastes in its definition of residues for substitution. The total HLW due for return is $1100m^3$ to which must be added an amount ($154m^3$) equivalent to 14 per cent HLW to be substituted for ILW/LLW that remains in the UK. He concludes that BNFL will only return $1254m^3$ (0.29 per cent) of the total volume of overseas fuel reprocessing wastes ($436\,000m^3$).

References

Berkhout, F. (1993), *Fuel Reprocessing at THORP: Profitability and Public Liabilities* (London: Greenpeace).

Black, D. (1984), 'Investigation of Possible Increased Incidence of Cancer in West Cumbria', Report of the Independent Advisory Group chaired by Sir Douglas Black (London: HMSO).

Blowers, A. (1984), *Something in the Air: Corporate Power and the Environment* (London: Harper & Row).

Blowers, A. and Leroy, P. (1994), 'Power, Politics and Environmental Inequality: A Theoretical and Empirical Analysis of the Process of "Peripheralisation"', *Environmental Politics*, 3, 197–228.

Blowers, A., Lowry, D. and Solomon, B. (1991), *The International Politics of Nuclear Waste* (London: Macmillan).

BNFL (1992), 'Radioactive Waste Management Advisory Committee: Submission on Waste Substitution', Doc Id 7008G.

BNFL (1993), *THORP: The Need, the Benefit, the Facts.*

Campaign for Freedom of Information, (1993) *The Environmental Information Regulations and THORP.*

Crenson, M. (1971), *The Un-Politics of Air Pollution: A Study of Non-Decisionmaking in the Cities* (Baltimore, MD: The Johns Hopkins Press).

Cumbria County Council (1992), *Review of Geology and Hydrogeology of Sellafield* (Environmental Consultants Limited).

Cutter, S. (1993), *Living with Risk* (London: Edward Arnold).

DoE (1986), Radioactive Waste (Professional Division) of the DoE, *Assessment of Best Practicable Environmental Options (BPEOs) of Management of Low and Intermediate-Level Solid Radioactive Wastes* (London: DoE).

DoE (1993), *BNFL Sellafield Further Public Consultation: Summary of Responses* (London: DoE).

Friends of the Earth (1992), *The MOX Myth, Special Briefing.*

Friends of the Earth (1993), *Discredited and Out of Date – The Government's Justification of THORP*; Friends of the Earth's Submission to the DoE's Consultation on the need for a THORP at Sellafield.

Gittus, J. (1992), *Public Attitudes to the Management of Irradiated Fuel'*, Institute of Mechanical Engineers, C447/057.

Government Statement (1993), *Statement of Government Policy on Reprocessing and Operation of the Thermal Oxide Reprocessing Plant at Sellafield* (London: HMSO).

Greenpeace (1993), *The THORP Problem – The Insoluble in Pursuit of the Incomprehensible*, THORP Papers 3.

Hickman, R. (1993), *Report of the Public Local Inquiry into Objections to the Proposed Spent Fuel Store at Torness Power Station, Dunbar, East Lothian.*

HMIP (1993), *Report on the Public Consulation Conducted by Her Majesty's Inspectorate of Pollution and the Ministry of Agriculture, Fisheries and Food (The Authorising Department) on British Nuclear Fuels PLC's Application for Revision of the Certificates of Authorisation to Discharge Liquid and Gaseous Low Level Radioactive Wastes and for a New Authorisation to Dispose of Cumbustible Waste Oil from the Sellafield Works at Seascale in Cumbria.*

HMSO (1984), *Disposal Facilities on Land for Low and Intermediate-Level Radioactive Wastes: Principles for the Protection for the Human Environment* (London: HMSO).

ICRP (1977), *Recommendations of the ICRP*, Publication 1.

Large and Associates (1992), *Comparison of the Radioactive Waste Arisings Generated by Reprocessing, Encapsulation and Storage of LWR and AGR Irradiated Fuels* (Greenpeace).

Large and Associates (1993), *Current Stocks and Future Arisings of Radioactive Waste in the United Kingdom: Contribution of Reprocessing Overseas Irradiated Fuels by BNFL from the Thermal Oxide Reprocessing Plant (THORP)* (Greenpeace).

Lindblom, C. (1977), *Politics and Markets: The World's Political-Economic Systems* (New York: Basic Books).

Loeb, P. (1986), *Nuclear Culture: Living and Working in the World's Largest Atomic Complex* (Philadelphia: New Society Publishers)

G. MacKerron, Surrey, J. and Thomas, S. (1994), *UK Nuclear Decommisssioning Policy: Time for Decision* (University of Sussex: SPRU).

NRPB (1992), 'Broad Statement on Radiological Protection Objectives for the Land-Based Disposal of Radioactive Wastes', NRPB Document, Vol. 3.

NIREX (1987), *The Way Forward: The Development of a Repository for the Disposal of Low and Intermediate-Level Radioactive Waste* (Harwell).

NIREX (1991), *The Repository Project – An Engineering Progress Report Describing the Preferred Design Concept* (Harwell).

NIREX (1993), 'Scientific Update 1993, NIREX Deep Waste Repository Project', *NIREX* Report No. 525, Harwell.

Radioactive Substances Act (1993), Decision by the Secretary of State for the Environment and the Minister of Agriculture, Fisheries and Food in Respect of an Application from British Nuclear Fuels for Authorisation to Discharge Radioactive Wastes from the Sellafield Site.

RWMAC (1990), *Eleventh Report* (London: HMSO).

RWMAC (1992a), *Advice to the Secretary of State on British Nuclear Fuels Plc Proposal for the Return of Waste Resulting from Reprocessing of Nuclear Spent Fuel to Overseas Utilities* (London: HMSO)

RWMAC (1992b), *Response to UK Nirex Limited's Revised Design for the Sellafield Repository* (London: HMSO)

RWMAC (1993), *Thirteenth Report* (London: HMSO).

Rosa, E. and Freudenberg, W. (1992), 'The Historical Development of Public Reactions to Nuclear Power: Implications for Nuclear Waste Policy', in R. Dunlap, M. Kraft and E. Rosa, *The Public and Nuclear Waste: Citizens' Views of Repository Siting* (Durham, NC: Duke University Press).

Royal Commission on Environmental Pollution (1976), *Nuclear Power and the Environment*, Sixth Report of the Royal Commission on Environmental Pollution (London: HMSO) Cmnd 6618.

UN (1992), *Agenda 21* (New York, United Nations).

Wolsink, M. (1993), 'Entanglement of Interests and Motives in Facility Siting: The not-in-my-backyard "Theory"'; Draft Paper.

Wynne, E. B., Waterton, C. and Grove-White, R. (1993), *Public Perceptions and the Nuclear Industry in West Cumbria* (Lancaster University: Centre for the Study of Environmental Change).

Zonabend, F. (1993), *The Nuclear Peninsula* (Cambridge: Cambridge University Press).

14 Running Up the Down Escalator: Developments in British Wildlife Policies after Mrs Thatcher's 1988 Speeches

Stephen C. Young

INTRODUCTION

This chapter examines issues relating to wildlife in Great Britain during the period between Mrs Thatcher's 1988 speeches on the environment and the government's response to the Rio Summit of 1992,[1] which came in January 1994 when the DoE published four reports on sustainable development and related issues (DoE, 1994a, 1994b, 1994c, 1994d). The argument of the chapter is that Thatcher's speeches had little significance in the sphere of wildlife conservation. The period from September 1988 through to January 1994 can best be seen as the second half of a period of change that began in the mid-1980s. The publication of the biodiversity action plan provides a natural cut-off point (DoE, 1994c). Before analysing the pre-1988 developments, and the issues relating to the institutional and policy changes of the 1988–94 period, three contextual features need setting out.

Laws Affecting Wildlife

The legal framework pertaining to the 1988–94 period was limited in its approach. It was established by the 1981 Wildlife and Countryside Act (W&CA). This consolidated and developed the wildlife aspects of the 1949 National Parks and Access to the Countryside Act and the 1968 Countryside Act. It also incorporated the 1979 EC Birds Directive.

   The wildlife protection policies that emerged were based on two approaches: the protection of significant areas, and the protection of rare and endangered species wherever they occurred. The area dimension was based on the declaration of SSSIs. The complex procedures for designating SSSIs were amended by the 1981 Act and by the 1985 Wildlife and Countryside (Amendment) Act. The criteria for judging whether an area should be designated as an SSSI were scientific. The aim was to protect and manage sites for the benefit of nature conservation. Criteria were developed to identify habitats that were unusual in national terms, like limestone pavements, or ones that supported good populations of rare species, like natterjack toads. The second approach was to identify endangered flora and fauna – such as polecats or rare orchids – and to protect them wherever they occurred, whether on an SSSI or not. Prosecutions are brought against people who damage such species or their habitats.

Government Attitudes towards Wildlife Issues

The key point to draw out here is that during the 1981–94 period wildlife protection policies frequently clashed with other government priorities. The government's underlying attitude was that policies on agriculture, transport, energy and industry had a core significance, while wildlife was of peripheral interest. Policies on wildlife were adapted to fit in with other priorities.

What emerged was that, from the government's perspective, wildlife issues could be divided into two types. First, there were the cases where it faced protests about wildlife that *challenged* its wide economic priorities – like pushing ahead with the roads programme. Second, there were other situations where pressures could be *easily accommodated* because they did not clash with other priorities. An example here is adding to the schedules of species protected under the W&CA.

This contrast between the ' *challenge issues*' and the ' *easily accommodated issues*' recurs throughout this chapter. It reflects a deep ambiguity in government attitudes towards wildlife issues.

Differing Attitudes among Conservation Groups on Wildlife Issues

It is misleading to speak of a wildlife conservation policy community or issue network. There appear to be – following Marsh and Rhodes (1991) – a range of the more closed policy communities and the more open issue networks. Detailed analysis shows that different groups have different perspectives and priorities on wildlife issues (Lowe and Goyder, 1983; Cox, Lowe and Winter, 1986: 181–6). Groups espousing more scientific approaches want to protect sites from people as well as developers. Landowning interests sometimes use wildlife arguments to preserve their estates. On the other hand, some of the urban wildlife groups have a more populist approach, aiming at drawing visitors on to interesting sites. Also wildlife groups sometimes become part of the wider rural conservation lobby trying to oppose development. There are further complications when groups line up on opposite sides at a public inquiry. During the 1979–94 era groups interested in wildlife were often able to unite against landowners or legislative proposals, but their differences did sometimes surface.

FEATURES OF THE 1979–88 PERIOD

First, the thrust of the Thatcher policies on the economy and planning prior to her environment speeches in 1988 were unsympathetic to wildlife conservation. The deregulation strategy to slim down the statutory planning system in the early and mid-1980s facilitated development (Thornley, 1991). The high-water mark of liberalising the planning system in rural areas came in 1987 – after the 1985 White Paper, *Lifting the Burden* (DoE, 1985) – with the reform of the Use Class Order (Marsden *et al.*, 1993: 122). Housing schemes, industrial estates, out of town shopping centres, enhanced leisure opportunities, infrastructure projects and other developments were all part of the 1980s boom (Buckley, 1989; O'Riordan, 1989; *Ecos* special issues on leisure, 1991, Vol. 12(1): 7–55, and coasts, Vol. 12(2): 1–22; Flynn and Lowe, 1992; Tyler, 1992). Important wildlife sites like Felixstowe Docks were damaged. Wildlife was also harmed by the long-standing inability of the post-war planning system to control agricultural intensification (Dixon, 1992; Evans, 1992).

Second, the W&CA proved weak and often ineffective (Pye-Smith and Rose, 1984). Under the legislation it was necessary to go through expensive, lengthy and cumbersome procedures to declare SSSIs, and then to manage them. The Somerset levels and Duich Moss on Islay are two of the many examples that illustrate the weakness of the Act (Lowe *et al.*, 1986). The NCC was the statutory body charged with carrying out the wildlife protection aspects of the W&CA and implementing the whole SSSI system throughout the UK. It tried to stand its ground, arguing, for example, the need to designate more sites under the 1971 Ramsar Convention, which protects sites of importance to waterfowl. In 1983/4 the NCC suggested adding 110 sites to the 19 that had been declared (Evans, 1992: 151). Derek Ratcliffe – the NCC's Chief Scientist in the 1980s – has shown how the NCC was marginalised (1989).

The third feature of the period was the backlash that the implementation of the W&CA provoked. The Act left the NCC having to regulate farmers through a horrendously complex piece of legislation (Brotherton, 1989;

Ecos, 1992, Vol. 13(2): 49). This provoked anger, especially in peripheral areas, and led to political pressures on a deregulating government to close down the NCC.

Changes in Agricultural Policy

However, despite all the anti-wildlife pressures, one trend emerged which was going in the opposite direction. By the mid-1980s agriculture was 'increasingly beleaguered by attacks on its budgetary extravagance and environmental insensitivity' (Cox, *et al.*, 1986: 211). Since the late 1940s it had been assumed that if the most significant wildlife sites were protected, then a healthy agricultural sector would benefit wildlife in the rest of the countryside. Conservationists gained support as they criticised this assumption from the 1950s through to the aftermath of the 1981 W&CA (Carson, 1962; Pye-Smith and Rose, 1984). Further, the Conservatives' drive to cut spending meant that the subsidies that were producing food mountains under the CAP were bound to be a target.

The imposition of milk quotas in 1984 marked the watershed in the moves away from the post-war approach of maximising food production. It was partly linked to attempts to curb CAP subsidies. The 1986 Agriculture Act redefined the aims of farming policy, giving greater prominence to conservation and other rural issues (Rydin, 1993: 130). The first evidence of wildlife issues getting a greater priority came with the ESAs scheme which was established by the 1986 Agriculture Act in response to pressures from the Countryside Commission as well as environmental groups. It drew from the experimental Broads Grazing Marshes Conservation Scheme (Colman and Lee, 1988). The aim of the ESA scheme is to identify areas of special landscape, wildlife or historic interest which are thought to be liable to damage from agri-business approaches. Payments are made to farmers who agree to farm in more traditional and environmentally friendly ways. For example, farmers in the Swaledale hay meadows do not cut the hay until after a specific date so that the wildflowers have time to seed.

Similar shifts took place in forestry policy. Conservationists had long been critical of the net loss of habitat from upland planting. The 1985 Wildlife and Countryside (Amendment) Act redefined the Forestry Commission's purpose, instructing it to give greater priority to wildlife and other concerns (Rydin, 1993: 134). This approach found expression in the 1985 Broadleaved Woodland Grant Scheme and the 1988 Woodland Grant Scheme. Higher grants were paid for the planting of native species. The argument here is that such trees support a greater diversity of insect

life, which provides an enhanced food source for birds. Also, tax concessions to private forestry firms were withdrawn in the March 1988 budget,
before Mrs Thatcher's speeches.

The policy initiatives in agriculture and forestry had implications at the
Quasi-Governmental Agency (QGA) level below MAFF. Both ADAS –
the Agriculture Development Advisory Service (Clarke and O'Riordan,
1989) – and the Forestry Commission (Turner, 1993) had to redefine their
roles. The changes encouraged farmers to look for new sources of income.
Some became more sympathetic to conservation issues.

Trends in Urban Areas

Much of the writing on urban areas in the 1980s focuses on the decline of
smokestack industries and on inner city problems (Robson, 1989).
However, this overlooks the significant initiatives that were taking place
on the wildlife front (Nicholson-Lord 1987; Smyth, 1987). Goode argues
that a new philosophy and a new approach emerged during the 1980s
(Goode, 1989). Instead of using scientific criteria to pick SSSIs the
emphasis was laid on sites that were unassuming. Even if they were of
limited value in national terms, they were significant in *local* terms for
two reasons: first, they were open green places where people could enjoy
themselves: secondly, teachers could take school children there to explore
the woods or go pond-dipping.

This new philosophy affected different kinds of sites. There was still the
need to protect the remnants of rare habitats. But it was also possible to
enhance and improve the value of JIMBOB PLACES for wildlife. These
are the places where people, children and dogs get up to all sorts of things,
such as Jogging, Insect-hunting, Mountain Biking, Ornithology, Black-
berrying, Picnicking, Lolloping around, Angling, Courting, Exploring and
Sun-bathing. What was really new in urban areas, although it was ana-
thema to some purists, was the idea of actually creating sites of wildlife
value. Criteria were developed to identify 'Areas of Deficiency'.
Ecologists and local people then worked out where pocket parks (Rose,
1990), wildflower gardens and ecology parks could be created. Often they
were in the corner of school grounds. Another approach was to identify
corridors along which mammals and birds could move into the city. Gaps
could be identified so that canals, railways, JIMBOB PLACES and formal
parks could be linked together for the benefit of local wildlife. Five features of this approach stand out from Goode's discussion. These are contrasted in Table14.1 with the ways in which similar issues were handled in
connection with the nationally significant SSSIs from the mid-1980s on.

Table 14.1   Features of protected wildlife sites, mid-1980s–mid-1990s

	Nationally important wildlife sites (SSSIs)	Significant wildlife sites in urban areas
Criteria for picking sites	Scientific criteria used to identify sites which will protect rare and endangered species and habitats	Scientifically unassuming sites are picked which, if protected, can give enjoyment to local people, and be of educational value for school children. The identification of 'Areas of Deficiency'
Policy aim and management regime	Aim is to protect and manage SSSIs so they retain their existing features	Aim is to enhance existing sites and even create new ones, through ecology parks, wildlife gardens, pocket parks,and wildlife corridors
The people involved in identifying sites	Highly trained specialists identifying sites	Trained and untrained people identifying potential sites
The people involved in managing sites	Experts working out management plans with landowners	Local community groups centrally involved in improving and managing sites. Not-for-profit partnership organisations often created (pp. 243–4).
Access	A permit system is used to limit access to those with a professional interest (e.g. PhD work, photography)	Access is encouraged for enjoyment and recreation; and for study in connection with the National Curriculum
Links to statutory plans	Policies for protecting these sites are worked into statutory planning documents	

Between 1984 and 1986 nature conservation strategies were published covering such authorities as the West Midlands County Council, the Greater London Council, and Greater Manchester County Council. Councils like Leicester (Lomax, 1990) and Edinburgh were developing similar ideas. The NCC was active in encouraging it once it had started. This activity was well established in urban areas before Mrs Thatcher's

1988 speeches and was spilling over into more rural areas. It symbolised a new populist approach to wildlife which was quite different from more elitist approaches based on scientists keeping the public away.

Not-for-Profit Organisations

Below the levels of Whitehall, QGAs, and local government there is a range of other organisations. They are involved in two distinctly different kinds of activity. There is the pressure group activity, as exemplified by the CPRE and the World Wide Fund for Nature (WWF). The latter lobbies within Britain, as well on an international basis. Second, there is the process of a range of organisatiosns running projects on a not-for-profit basis. For example, the RSPB purchases land and manages nature reserves and the National Trust not only acquires buildings, but manages land with wildlife considerations as a priority. This is a long-standing tradition, reaching back to the Victorian era (Lowe, 1983; Evans, 1992). Britain is unusual in having a range of these private, non-public sector organisations acting in a semi-official capacity.

There is an enormous variety of wildlife organisations running projects on a not-for-profit basis. Some, like the RSPB and the National Trust, are long established. The County Wildlife Trusts (CWTs) and their umbrella organisation, the RSNC – The Wildlife Trusts Partnership, have evolved from bodies that were established from 1912 onwards (Smith, 1990). Others, like the urban wildlife trusts that emerged in the 1980s, are fairly new (Millward, 1990). Some, like the Groundwork Trusts, were created as a result of a top-down government initiative. On the other hand, local groups in urban areas, protesting about the loss of their green space, have evolved into not-for-profit trusts to improve and maintain their wildlife area. Such organisations have emerged from the bottom-up, from communities and neighbourhoods.

These not-for-profit organisations cover a variety of geographical areas. The community-based ones have just been mentioned. The British Trust for Conservation Volunteers (BTCV) and the Wildfowl and Wetlands Trust on the other hand, cover the country. The Woodland Trust started as a regional body in the South-West and evolved into a national organisation. CWTs relate to local authority boundaries.

The numbers of organisations involved in running projects on a not-for-profit basis grew substantially during the 1980s. Some are involved primarily in conventional pressure group activity aimed at changing policies at central, QGA and local government levels. Growing numbers focus more specifically on running projects. The significant point here is that the

1980s saw an expansion of this activity both in terms of the numbers of organisations, and in the budgets of individual organisations.

INSTITUTIONAL CHANGES 1988–94[2]

It is clear that by the time Mrs Thatcher made her speeches in September 1988, a number of significant changes were already afoot. The speeches themselves came at an awkward point for Nicholas Ridley, the Secretary of State for the Environment. Always one who thought he should do the right thing rather than be popular, he was planning to dismember the NCC. Just before his resignation in July 1989 he proposed splitting the NCC into separate organisations for England, Scotland and Wales, and amalgamating the Countryside Commission arms for Scotland and Wales with the relevant parts of the NCC. In England, the Countryside Commission and the remnant of the NCC would remain separate. He justified this on the grounds that the NCC's nationwide approach was not merely inefficient, but insensitive to local needs. Critics suspected that this was Ridley's revenge on an organisation that had tried to stand up for conservation interests. Ridley and the Secretaries of State for Wales and Scotland had been impatient with its ability to delay economic development projects. In Scotland, the Highlands and Islands Development Board was very antagonistic, and the NCC's attempts to promote management agreements on SSSIs had also alienated the landowning lobby (*Ecos* special issue on Scotland, 1989, Vol. 10: 1–33).

An enormous row erupted into the media (Mayes and Smith, 1990; *British Wildlife*, 1990, 1: 365). Most conservationists argued that splitting up the NCC would undermine its scientific capability and its accumulated expertise. In Scotland and Wales though, some saw benefits in having locally based organisations dealing with landscape conservation as well as wildlife (Caldwell, 1993; Pepper, 1993). As new legislation was required, the row became protracted. This damaged the credibility of the new Secretary of State for the Environment, Chris Patten, as he started work on the environment White Paper (DoE, 1990) in September 1989. The episode also tarnished Mrs Thatcher's new environmental image. More specifically, from the perspective of this chapter it damaged relations between the Government and conservation interests at a time when it had looked as if they might improve substantially.

Eventually the changes were enacted through the EPA and the 1991 Natural Heritage (Scotland) Act. The new organisations began work in April 1991: English Nature (EN), Scottish Natural Heritage (SNH), and the Countryside Council for Wales (CCW). Government concessions led to money being given to the new agencies, and to the creation of a Joint Nature Conservation Committee to coordinate the work of the three bodies and promote international liaison. Early in 1994 the DoE proposed merging EN and the Countryside Commission to create one organisation in England.

In 1991 the Forestry Commission was split into two (*British Wildlife* 2: 252–3; Rydin, 1993: 134). The Forestry Authority regulates all forestry, carries out research and administers the grants. Forestry Enterprise is responsible for planting and management. This split was partly in response to complaints from wildlife groups like WWF and the RSPB. Despite the amendments to forestry policy in the 1980s, the damage to wildlife continued (Ratcliffe, 1993). Apart from destroying moorland, upland plantations double the effects of acid rain. This not only kills fish and insects in rivers, but has knock-on effects, as on dipper populations in Wales. Environmental representatives on the regional advisory committees are meant to balance commercial interests. Also in 1991, the Water Resources Act imposed on the NRA the responsibility of carrying out its functions with regard to wider considerations, including the conservation of flora and fauna.

There were no significant institutional changes at the level of local government during the 1988–94 period, apart from the development of Green Forums. Urban authorities like Calderdale and Islington created these forums to draw conservation expertise into their decision-making processes. This trend also affected sensitive sites like the Burry Inlet. However, in the not-for-profit world new organisations continued to emerge. The number of Groundwork Trusts increased to 30. Plantlife was launched in 1989 to protect threatened British plant species, and to do for plants what the RSPB had done for birds (*British Wildlife*, 1: 53/4; and 1: 110).

One significant development amongst the CWTs was the setting-up of consultancy arms to help earn income (Shirely and Knightsbridge, 1992). They had suffered from the loss of the Manpower Services Commission money that had been available in the 1980s. The Lancashire and Avon Wildlife Trusts were amongst those that set up arms' length commercial operations. They aimed to create earning opportunities, a more stable environment and greater autonomy.

THE EVOLUTION OF POLICIES 1988–94

The period saw a move away from the NCC's regulatory style to a partnership approach. In mid-1992 EN announced it was changing the financial basis of its policy towards protecting SSSIs (*British Wildlife*, 1992, 4: 61). Previously the NCC had operated on the basis of ' profits forgone'. Officers worked with landowners to devise management arrangements that helped protect the integrity of the site from a nature conservation perspective. But where owners of SSSIs succeeded in claiming they had to forgo profits by, for example, not proceeding with forestry projects, they were compensated for the profits forgone (*British Wildlife*, 1991, 3: 56). EN decided it would no longer offer payments on a profits-forgone basis; instead it developed positive management agreements with SSSI owners, drawing from its experiments with wildlife enhancement schemes. The number of agreements jumped from 1053 in 1986 to 1759 by April 1990 (Adams, 1991). Islay provides a good example of the changing style. Farmers greatly resented the old NCC's arrogance in announcing SSSIs in the early 1980s without consultation, yet a decade later SNH had a local office and was cooperating with farmers and compensating them for damage done to grassland by wintering geese.

An important rethink of principles and practice also took place with regard to the wider countryside away from SSSIs (Colman *et al.*, 1992 and 1993). EN developed its Natural Areas approach. This divided the land up by geological features and the types of habitat they supported.

Part of the rethink involved experiments with economic valuation techniques (Pearce *et al.*, 1989). The 1991 review of ESAs led on to the Countryside Stewardship scheme in England, and Tir Cymen in Wales. These drew from the experience of operating management agreements and incentives on SSSIs and ESAs in the late 1980s. The main feature of the Stewardship and Tir Cymen approaches is that they are market-led. They are exploring the extent to which farmers can be induced to produce – at a set price – environmental goods of benefit to society. These include permissive paths, open access arrangements, and, on the wildlife front, the recreation and management of habitats like orchards and heather moorland (Dixon, 1992; DoE, 1992: 80–1; Bishop and Philips, 1993).

The wider post-Brundtland debates about sustainable development were also influential (Stoker and Young, 1993: ch. 4). EN experimented with the managed retreat of coastlines, trying to recreate saltmarsh of the kind that would be drowned by rising sea-levels (Maguire and Barkham, 1991).

In January 1994 a lot of the new thinking was drawn together with the publication of *Biodiversity: The UK Action Plan* (DoE, 1994c). This tried to give practical expression to the government's signing of the Convention on Biological Diversity at the 1992 Rio Summit. Although it was designed to generate its own targets, it was criticised by conservationists for its ill-defined aims, and the lack of commitment and resources needed to achieve its aims (Avery, 1993; *British Wildlife*, 1994, 5: 198–9). Six conservation groups published an alternative plan to show how a much more ambitious document could have been published (Butterfly Conservation *et al.*, 1993). *Sustainable Forestry: The UK Programme* (DoE, 1994d) – also published in January 1994 – focused mainly on management approaches. However, it had implications for wildlife, as with the proposals to expand tree cover through community forests.

Another innovation was the NRA developing a much more aggressive prosecuting style towards polluters than its predecessors' approach. In the 1989–93 period it secured prosecution in 548 out of the worst 1776 – Category 1 – pollution incidents (51 have yet to be resolved). Although it was often criticised for only prosecuting in 2 per cent of all reported incidents, it was successful in about a third of the major ones (Young, 1993: 108).

Challenge Issues and Easily Accommodated Issues

Beyond policy innovations, it is necessary to distinguish between the two different kinds of policy identified at the start of this chapter. With the *easily accommodated* issues it was relatively straightforward for government to yield to accommodate conservation pressures. The significance of Mrs Thatcher's speeches here was in speeding up the incremental development of policies on a range of issues. For example, the 1991 review of ESAs led to an expansion of the scheme (Dixon, 1992).

However, on topics that *challenged* government priorities up to January 1994, ministers ignored group demands, forging ahead with schemes designed to promote economic development. Typical examples include the review of SSSI arrangements in Scotland (*British Wildlife*, 1992, 3: 386); and the roads programme (*Ecos* special issue on transport, 1992, 13(4): 1–22; *British Wildlife*, 1992, 4(1): 65, and 4(2): 132; Young, 1994a).

Conservation interests did score some isolated successes on the *challenge* issues. In 1991, the DoE ordered Poole Council to revoke planning permissions at Canford Heath, although much damage had already been done (Haskins, 1991; *British Wildlife*, 1991, 2: (254). Also the Department of Transport (DoT) withdrew the Hereford by-pass proposal in part because of the inspector's criticism of the damage it would do to ancient meadows (Woodford and Oliver, 1992).

On the whole, policies were based on the voluntary principle (Cox, Lowe and Winter, 1990). Bureaucratic controls were sometimes used, as with the regulations being backed by fines for damaging flora and fauna listed in the schedules relating to the 1981 W&CA. But the emphasis was on financial instruments and voluntary cooperation (Pearce *et al.*, 1989; Young, 1993: ch. 6). These were used to attract the voluntary cooperation of landowners as over SSSIs, ESAs, farm woodland programmes, and the Nitrate Sensitive Areas Scheme.

THE IMPACT OF GOVERNMENT 1988–94[3]

Some initiatives led to positive gains. EN developed a Species Recovery Programme worth £250 000 in 1992/3 (*British Wildlife* 1992, 3: 251, 1993, 4: 332; and Marren and Rich, 1993). The re-introduction programmes bringing sea eagles and red kites from Scandinavia to Britain continued. Also of significance were the attempts (discussed earlier) to operationalise sustainable development.

There was growing appreciation of the need to protect sites away from the SSSIs. In cities this meant the JIMBOB PLACES referred to earlier, such as Nover Hill in Bristol (*British Wildlife*, 1993, 4: 337). By 1994 though, such sites still lacked a statutory designation equivalent to a Tree Preservation Order. Other gains at the local level were the increased momentum in the continuing publication of wildlife strategies and the promotion of wildlife corridors (Tyldesley and Collis, 1990).

The next positive achievement of the 1988–94 period was the broadening of the information base. Comprehensive surveys provide baseline data, making it easier to analyse changes, prioritise programmes, and assess the significance of proposed development sites for wildlife. A typical example is the British Trust for Ornithology's new edition of its 1976 atlas of breeding birds (Gibbons *et al.*, 1993). Also, research work led to new approaches to best practice. The Game Conservancy's work on conservation headlands on arable farms, for example, showed how wildlife could be promoted around the edges of fields. In a number of cases groups tried to identify implementation problems with existing policies. This monitoring picks up problems which can then be corrected via a policy implementation/feedback loop (Barrett and Fudge, 1981). An example is research on how to improve the value of set-aside land for wildlife (Wilson and Fuller, 1992).

Government can claim credit for some initiatives, but not for making concessions under pressure, as over Oxleas Wood. It was more positive over the collapse of conifer planting and the expansion of broad-leaved planting, even though it was responding to pressure. Also some gains came partly because of help from not-for-profit organisations. WWF provided half the funding for the Species Recovery Programme, for example.

The 1988–94 period also produced negative results. On the traditional yardstick of measuring conservation policy – damage to SSSIs – Britain went backwards. Adams (1991) and Rowell (1991) suggest that at least 5 per cent of SSSIs were being damaged in any one year. It was claimed that about 7 per cent of England's SSSIs were damaged in 1992–3; and 20 per cent of Wales's (*British Wildlife*, 1994, 5: 266). The criticism that the powers to protect SSSIs were inadequate grew (Rowell, 1991; Evans, 1992). The lagoon sandworm is an example of a species that became extinct (*British Wildlife*, 1992, 3, 189).

Similarly, progress on designating Ramsar sites, Special Protection Areas, National Nature Reserves and Marine Nature Reserves was very slow. For conservationists, the lack of designations was one of the most worrying issues (Warren, 1991; Brown, 1992). The reason for it is simple. Once a site is an SPA there are more difficult procedures – with EU dimensions – to go through to put a motorway or a barrage in place. Holding designations back thus keeps Whitehall's options open, especially on *challenge* issues.

Finally, on the negative side of the evaluation, worrying long term trends continued to develop. Some trends are little understood, such as the ill health of seals and fish in the North Sea, while others are understood, but difficult to reverse, such as the stealing of limestone for rockeries. Even more worrying are the long term trends that are difficult to control. Acid rain will continue to fall on upland SSSIs in England and Wales for at least a decade (*British Wildlife*, 1993, 4: 331, and 4: 404).

Trying to Run up the Down Escalator: Towards an Assessment

What this all adds up to is not clear. By 1994 conservation interests had won some victories, and opened up new issues. But many groups remained depressed both by the scale of the tasks confronting them and by long-term trends not being amenable to government action. Making progress was like trying to run up a long escalator that goes down. You make some gains but, however fit you are, they are limited. You may glimpse what is at the top, but you usually end up exhausted, going back down.

In some respects British conservation policies had, by 1994, moved *away* from maximising agricultural production and the negative approaches of the early 1980s. But it was not clear what they had moved *to*. ADAS and the Internal Drainage Boards had begun to change their spots (*British Wildlife*, 1994, 5: 201). Also by March 1994 there were 22 ESAs covering 10 per cent of agricultural land, but only 5 per cent of the projected 1995/6 spending of £2880m on the CAP related to agri-environment schemes (*British Wildlife*, 1994, 5: 264 and 266). In March 1994 the roads programme remained at £20 billion after cuts of only £3 billion. Also no clear conclusions about cost-effectiveness had emerged from the experiments with economic instruments (Potter, 1993), or about how to reconcile sustainable development ideas with wildlife conservation. Using barrages for renewable energy, for example, damages habitats.

By early 1994 there was little by which to judge EN, CCW and SNH, the new agencies (Caldwell, 1993; Pepper, 1993; and *Ecos*, 1993, 14(1): 65–6). Since their establishment in 1991 they had been largely preoccupied with internal restructuring and public relations. The focus of their debates and policy documents were mainly on principles and ideas. In terms of action CCW got furthest in trying to work out what ' partnership' meant, and how to integrate landscape policies with wildlife conservation concerns.

The 1988–94 period was thus one of transition, questioning assumptions and rethinking, as over the roads programme. A full assessment will depend on the impact of measures being developed during the period; on the implementation of *Biodiversity: The UK Action Plan*; and on more time elapsing so the balance between the conflicting trends can be evaluated. Also by early 1994 the issue of forestry privatisation had not been resolved, and the government's final response to the EC's 1991 Habitats Directive was still unclear (*Ecos*, 1993, 14(2): 55–6). By 1994 an air of confusion hung over the impact and the future of the conservation policies developed after 1988. The best that can be said is that they were, with the initial changes of the mid-1980s, a second step away from earlier policies and Ridley's ideas. But conservationists argue there is still a long journey ahead, which is made more difficult by having to run up the down escalator to get there.

INFLUENCES ON POLICY 1988–94

With the *easily accommodated* issues, the pre-1988 trends continued to exert an influence through the 1988–94 period. The new philosophy and approach in urban areas encouraged local authority and not-for-profit activity. Local authorities realised this was an area where they could develop some autonomy (Ward, 1993). Other examples were the Stewardship and Tir Cymen schemes growing out of the ESA programme and pressure from the CLA and others in 1987/8 (Bishop and Philips, 1993); and the slowing of publicly funded conifer planting (Turner, 1993; Yull, 1993).

In the background, international obligations also had an impact. On issues like peatland management techniques in Scotland, and changes associated with the ESA and CAP schemes, EC initiatives continued to nudge the development of policy along (Baldock, 1989). Although Britain ratified the Berne Convention in 1982, and the Bonn Convention in 1985, these had little impact. However the government did step up the designation of sites under the 1971 Ramsar Convention (Evans, 1992: 51–4).

Thatcher's 1988 speeches had little impact on conservation issues, partly because she ignored the comments of Heseltine and other Tories on conservation, and tried to carve out an international environmental agenda (Flynn and Lowe, 1992). The significance of Thatcher's speeches was that they helped to change the climate and speed up incremental change on the easily *accommodated* issues. But other factors also influenced this change of climate. McCormick (1991) and Robinson (1992) point, amongst other things, to the 1989 14.9 per cent Green Party vote in the European elections, interest in green consumerism, and disasters like the 1986 Chernobyl and 1989 Exxon Valdez cases.

This new climate was significant for two reasons. First, groups – which were becoming increasingly professionalised (Rawcliffe, 1992a) – were able to exploit it. Suddenly, after the rigours of the 1980s, the situation seemed more promising. Groups were active in four ways. First they had a prominent role in opposing government policies, as over the abolition of the NCC and continuing damage to SSSIs. Next they monitored government schemes. On ESAs, for example, they argued for revisions and extensions (Dixon, 1992). Third, they committed resources to commenting on draft government documents like the Planning Practice Guidance Notes (Burton, 1990). An important feature here was the way they pushed ideas about applying sustainable development (Owens, 1993). Finally, they moved on from trying to amend existing policies to highlight new issues, as over the need to protect estuaries (Rothwell and Housden, 1990). These four roles were also played at the local level by CWTs, local Friends of the Earth groups, scientific societies (Rotherham, 1990), and a range of conservation groups.

The climate was also significant on one major issue where the government changed tack. As noted earlier, one of the pre-1988 trends had been the deregulation of the planning system to help promote development, but this approach was reversed in 1989 by Patten. He initiated the *re*-regulation of rural planning (Burton, 1990; Marsden *et al.*, 1993: ch. 5). The government made concessions – which were partly to the benefit of wildlife – in response to 'strong environmental and conservationist pressures' (Marsden *et.al.*, 1993: 119). This had been a *challenge* issue, but

groups were able to exploit the climate of 1988/9. Here Thatcher's acceptance of Ridley's resignation was more significant than her speeches.

Moving on to the *challenge* issues, landowners continued to be
influential (*Ecos*, 1990, Vol. 11: 69). Ministers were committed to a position on Cardiff Bay which they defended determinedly (Harvard and
Ferns, 1993). On transport, although some anti-road campaigns were successful – Oxleas Wood for example – government was largely able to
ignore them during the 1988–94 period (Young, 1994a).

Three other factors came into play on the *challenge* issues. First, there
were inter-departmental tensions which led to policies being watered
down. A typical DoT example is the way it objected to the DoE declaring
the New Forest a National Park or an SPA, as this would have made
widening the A31 much more difficult. As agricultural corporatism (Cox
and Lowe, 1984) declined, MAFF searched for new functions and
justifications. This was also a source of conflict.

A second factor was the British courts. In 1990, WWF and the British
Herpetological Society took Poole Council to the High Court over the grant-
ing of planning permission for housing on parts of Canford Heath. This was
the first time the courts had been used to try to protect an SSSI. The site was
an important breeding area for protected species like the sand lizard, smooth
snake, Dartford warbler, nightjar and hobby. Although the case was lost, the
DoE revoked the planning permission in March 1991. This was partly as a
result of the hostile reception given to a DoE delegation to the Berne
Convention in Switzerland (*British Wildlife*, 1991, 2: 187/8, 1992, 3: 185/6).

The third factor capable of wounding the government on *challenge*
issues was the EC and the European Courts. There were arguments over
whether the Environmental Assessment directive had been properly
applied in Twyford Down and six other cases. In the end the EC stepped
back on these, though it may have been a factor in the government's decision to withdraw the Oxleas Wood proposal. The problem with most EC
interventions of this kind is that they occur after projects have com-
menced. For example, nearly half of an 800 hectare site in Kincardineshire
had been planted with conifers before the EC called a halt. The RSPB had
complained about the lack of an environmental assessment (*British
Wildlife*, 1992, 4: 62). However, EC intervention did help solve the Islay
Duich Moss case before damage was done.

The Diffused Nature of Policy-Making Processes on Wildlife Issues

The main problem for wildlife interests during the 1988–94 period was
the fragmentation of the state. Policies on aspects of energy (Ebrahim and

Elliott, 1991), transport, urban renewal and other issues affect wildlife. The number of QGAs and departments that were involved is larger than those specifically mentioned in this chapter. In the 1988–94 period, the consideration of wildlife issues in wider policy-making processes at national and local levels was erratic.

In theory, the DoE – and the Scottish Office and the Welsh Office in their areas – had responsibility not just for coordination of policy on wildlife issues, but for providing clear leadership. However, the fragmented nature of policy making processes, and the competing government priorities on challenge issues meant, that the leadership was timid and ambiguous, as, for example, over planning around estuaries (*Ecos*, 1991, 12(2): 1). Ridley's attempt to impose ideas about a contracting role for the state and privatised nature reserves (O'Riordan, 1989) had little impact beyond abolishing the NCC. He resigned as Secretary of State for the Environment in July 1989. Marsden *et al.* sum the situation up as ministers being 'torn between wishing to appease the conservation lobby while not offending landowning interests – and the result has been stalemate' (1993: 119). Wildlife is thus a complete contrast with areas like training (King, 1993), housing (Malpass and Marie, 1990), urban renewal (Healey *et al.*, 1992) and local government (Stoker, 1991; Allan *et al.*, 1994), where strong policies are set out by the centre and implemented through a variety of public, private and not-for-profit organisations.

Wildlife policy thus lacked overall coherence during the 1988–94 period. Together, weak leadership and the fragmented machinery encouraged drift. They reinforced the divide between the *challenge* issues, and those that could be relatively *easily accommodated*. This was the situation in which the conservation agencies like EN had to operate and to which wildlife lobbyists had to relate. The pro-conservation pressures were diffused.

There appeared to be – following Marsh and Rhodes (1991) – a range of policy communities and issue networks on different issues. The features of policy communities are restricted membership; stable relationships and strong dependencies between the organisations involved; and insulation from other networks, Parliament and the public. On peat (Rawcliffe, 1992b), red deer issues in Scotland, the management of the Cairngorms (Aitken, 1992), and roads (*Ecos* special issue on transport, 1992,/3(4): 1–2), conservation interests were excluded. At the local level, wildlife policy communities had been established during the 1970s and 1980s as in Leicester, by officers wanting to exploit the specialist knowledge of groups like the CWTs (Rydin, 1995). It seems that, after 1988, such communities were relatively untouched by the growth of group interest on a range of environmental issues.

Issue networks, on the other hand, are characterised by more fluid membership, permeable boundaries, and organisations with conflicting views and a lack of dependence on each other. The extensive consultation procedures over the preparation of the 1994 *Biodiversity* report involved many conservation groups (Avery, 1993). Other good examples are community forests (Johnston, 1991), and the more populist approaches to urban wildlife.

However, a paradox emerged. Intense debates and increased group activity post-1988 appeared to provide evidence of pluralism. Yet the extent to which policy making processes were opened up to wider issue networks was deceptive, and varied geographically too (Lowe *et al.*, 1993). Policy making on commercial forestry, for example, appeared to remain within a closed policy community despite widening public debates (Tomkins, 1989 and 1993; *Ecos*, 1990, 11(4): 71 and 1992, 13(2): 47–8; Yull, 1993). The confusion of the 1988–94 period is illustrated by MAFF reclaiming the ESA issue and closing up the debate about the nature of agri-environment policy after wide-ranging consultations over the future of the ESA programme in 1991. This has been seen as evidence of the resilience and reconstitution of agricultural corporatism (Marsden *et al.*, 1993: 63–4, 178–85).

This variety and complexity means that consensus and adversarial approaches co-existed (Rawcliffe, 1992a, 1992b). Where groups were excluded, this affected their campaigning style. The CPRE and the National Trust, for example, appear to be at the centre of the issue network on rural planning, yet on roads topics they were excluded from the policy community and became more unconventional, confrontational and uncompromising, even if not as radical as Earth First! (Young, 1993: 22–3). The National Trust threatened not to cooperate with the DoT over the road scheme at Hindhead and to force it to retreat to get a Parliamentary vote to approve the proposed A3 widening. The DoT gave in and produced the tunnel scheme it had said was too costly.

By 1993/4 a potentially significant new factor had emerged. Policy making processes involving environmental issues were being amended in Whitehall as a result of DoE interest in sustainable development, and the bilateral discussions it held with other departments while preparing the January 1994 documents. This led to environmental – and wildlife – issues being raised and considered in a more systematic, routine way in the context of wider decision making processes on transport, planning, and other issues (Young, 1994b). This change was a result of criticisms of the lack of integration of environmental issues into wider policy making processes (Sands, 1990; Hill and Jordan, 1993), and of DoE reactions to the 1992 Rio Summit, not Thatcher's 1988 speeches. The problem from a conservation perspective, however, was that wildlife issues did not carry much weight in the amended processes, especially on the *challenge* issues.

NOT-FOR-PROFIT ORGANISATIONS AS A THIRD FORCE

In the 1970s, conservation groups moved on from lobbying decision makers to courting media attention (Lowe and Goyder, 1983). One of the striking features of the 1980s was the way these groups moved on again to devote resources to carrying out conservation projects on their own. This was picked out as a long-standing pre-1988 trend in the earlier section on developments through the 1980s. It continued to develop during the 1988–94 period. Established organisations like the Wildfowl and Wetlands Trust, the RSPB, the National Trust, the Woodlands Trust and the CWTs at the local level increased their reserve coverage. New organisations like the canal trusts, Groundwork Trusts, and a variety of urban wildlife groups running their own projects continued to emerge.

This trend continued for four reasons. First, people involved in conservation issues in many different places were more and more often coming to the conclusion that government would not act on issues conservationists saw as important. As a result, groups moved on from trying to influence policy to planning and carrying out their own projects. They argued that the only safe sites were those owned by conservation bodies.

The second reason was a growing acceptance of the reality that there were limits to what both public and private sector organisations could do on the wildlife front. The government was ambivalent about the significance of wildlife, especially on the *challenge* issues discussed earlier. The public expenditure climate was increasingly hostile around the turn of the decade. Also, private companies cannot earn profits from wildlife. They can accommodate it as a sideline, but not as a priority. For example, foresters can cut woodland rides at the optimum width for certain butterflies, and similarly, developers in urban areas can contribute to wildlife corridors; but such activities are peripheral to the private sector's operations. The model of the not-for-profit organisation was ideally suited to move into the vacuum created by the inability of public and private sector bodies to respond to growing calls during the 1988–94 period to give conservation a higher priority.

The third reason why these organisations became more significant was that all three levels of government – central departments, QGAs and local councils – realised that they could make practical use of these organisations.

They could be used to channel resources to tackle specific problems. They could also shoulder long-term maintenance responsibilities.

Finally, state agencies learnt that public money could be made to go further when targeted at not-for-profit organisations. This is because they have access to a mix of resources not available to public sector bodies. They can get funds from all levels, from Brussels down to the parish council, and they can also access gifts in kind from private companies, and money to sponsor specific projects. Further, they can plug themselves into big charitable foundations such as Carnegie, and earn extensive income from membership fees. They thus provide a mechanism for extending the impact of limited public funds.

These not-for-profit organisations are best seen as public/private partnerships drawing resources together from the different sectors. As mentioned earlier, they are certainly not new. The not-for-profit model flowered during the 1980s and early 1990s, and became more significant. Although it has grown from traditional voluntarism, it needs to be distinguished from it.

These not-for-profit organisations are best understood as a breed of organisation that has become more common: a hybrid, a cross between a pressure group and a QGA (Stoker and Young, 1993: ch. 6). They are independent and outside government like a pressure group, but instead of focusing just on policy-making they get involved in implementation like a QGA. They have access to public sector funds like QGAs but they do not rely entirely on the public purse. They can raise funds from other sources like a pressure group. They are detached from Whitehall like a QGA, but like a pressure group, they are independent and not subject to ministerial direction. These organisations have a greater range of activity than a conventional pressure group which concentrates exclusively on lobbying. The point here is that they go beyond lobbying to run their own projects. It is the changing balance between these two activities that is significant.

Not-for-profit organisations operate across the increasingly hazy boundary between the public and private sectors. The term 'Third Force Organisation' (TFO) is useful as it conveys the way these organisations supplement the public and private sectors (Stoker and Young, 1993: ch. 6). The term 'voluntary sector' is often used (Petersen, 1990; J. Bishop and Rose, 1992). This may refer to many local wildlife organisations, but the not-for-profit partnership model that draws finance together from a range of sources fits lots of other organisations too from the National Trust to the Groundwork Trusts. Similarly, 'NGO' is conventionally interpreted as meaning pressure group (Grove-White, 1992: 11). However, this is loose and misleading, given the way they supplement their attempts to influence agendas with their growing involvement in specific projects. 'NGO' also has connotations of an organisation involved in North/South issues. Farrington *et al.* (1993: 3) argue that the all-inclusiveness of NGO makes the term 'almost meaningless'. They sum the position up as follows: 'There is considerable confusion in both the literature and among policy-makers as to what we mean by the label NGO.'

TFO provides a term that covers organisations running projects on a not-for-profit basis that emerge locally or at regional or national levels from people with similar ideas in civil society. Examples like those discussed earlier can be said to be *bottom-up* TFOs. On the other hand, TFO also covers those that operate on a not-for-profit basis which have been set up by government in *a top-down way*, like the Groundworks.

The way to place this increasingly common breed of organisation on the political map is to understand it in terms of the blurred boundary between the public and private sectors. In the 1960s and 1970s it was fairly clear where the public and private sectors joined. TFOs are partnership organisations drawing resources together from the two sectors. As a result, the public and private sectors began to overlap. The whole process can best be explained by the concept of a mixed sector emerging between the two sectors (Young, 1992). The mixed sector is the arena within which partnerships operate. They span the traditional boundary between the public and private sectors. The concept of the mixed sector is set out in Figure 14.1. It is analogous to the position where the sea meets the land: there is no one point; there is high tide and low tide. Between those two points there is the inter-tidal zone with an ecology all of its own: the salt marshes, the mudflats, the rock pools and so on. An enormous variety of flora and fauna exist there, and creatures from other parts of the planet live off them. Similarly the mixed sector is home to a variety of organisations which are exploited by public and private sector bodies.

This concept of a mixed sector is paralleled in other policy areas. In urban renewal, for example, there is a wide variety of TFOs. In that area there is also another kind of partnership. Figure 14.1 distinguishes between not-for-profit partnerships of the TFO kind, and profit-oriented partnerships. In inner city areas where prospects for profits are uncertain, private sector companies entered into a whole range of partnerships with QGAs and local councils during the 1980s. These urban renewal projects involved some kind of public subsidy to attract private sector investment. This reduced the net cost to firms, making it possible for them to develop projects on unlikely sites and earn conventional profits (Healey *et al.*, 1992). The economics of wildlife are such that profit-oriented projects are not usually available. Profit-oriented partnerships have thus not been included in this chapter.

CONCLUSION

In the late nineteenth and early twentieth centuries, organisations such as the National Trust, the RSPB and the RSNC's forerunner, the Society for the Promotion of Nature Reserves, tried to act on their own. Over the years they and others encouraged the state to expand its responsibilities. As in the social services, the experiments of voluntary organisations were subsequently taken on by the state (Brenton, 1985).

The 1980s, and in particular the 1988–94 period, can best be seen as the approach to a new watershed. The central state was trying to limit its activities, divesting itself of responsibilities, and drawing other actors into the front line. It used financial instruments to secure private sector cooperation. It also tried to apply the enabling authority approach in local government (Brooke, 1989; Stoker, 1991) to wildlife, and to explore the potential of TFOs. The 1988–94 period was one of continuing vitality for TFOs as they grew in numbers and responsibilities, taking on some of what the state used to do. Whitehall's strategy was to try to redefine the roles of the three levels of government – central, QGA and local – and to work out what could be left to those in civil society–landowners, firms and TFOs.

However, no new consensus emerged about the roles of different actors. The government was widely criticised for shirking its responsibilities and for denying TFOs the resources they needed to do the jobs they were being given.

Some argue that, by early in the twenty-first century, the state's roles will be limited, and that there will be a return to the pattern of earlier in the twentieth century. Others suggest that in time, the state's role will expand, and that it will take on some of the responsibilities being developed by TFOs. In respect of the state's roles, the politics of wildlife parallels debates in the spheres mentioned before, such as housing, urban renewal, and local government. It is not clear how far the Thatcher and Major premierships have cut back the state's role in the longer term, or whether the post-war pattern of governments shouldering increased responsibilities will reassert itself. Weak leadership during the 1988–94 era prolonged what was essentially for wildlife policy, a period of transition: a search for new conventions about the roles of different actors in the political system.

Notes

1.   I am most grateful to the participants in the Newcastle Colloquium for their reactions to the initial draft of this paper, and in particular to Philip Lowe

and Steven Yearley. Susan Baker and David Colman also provided stimulating comments. However, I do of course remain responsible for what is here.
2. The rest of this chapter draws extensively on the empirical side from the summaries of events in the 'Conservation News' section of *British Wildlife*. Sue Everett became the regular writer of this section from August 1991 when Vol. 2 (6) was published. Where I have drawn from her insights I have indicated this by reference to *British Wildlife* in the text.
3. There is no space here to draw in Britain's overseas contributions to global biodiversity problems: for a survey see DoE (1994: ch. 8). NI issues are discussed in a separate chapter in this volume.

References

Adams, B. (1991), 'SSSIs: Who Cares?', *Ecos*, 12(1), 59–64.
Aitken, B. (1992), 'The Cairngorms – Still More Paper, Still No Action', *Ecos*, 13(4), 54.
Allan, P. *et al.* (1994), *Focus on Britain 1994* (Deddington: P. Allan).
Avery, M. (1993), 'Biodiversity: The UK Action Plan', *Ecos*, 14(3/4), 70–1.
Baldock, D. (1989), 'The EC and Conservation in the Thatcher Decade', *Ecos*, 10(4), 33–7.
Barrett, S. and Fudge, C. (1981), *Policy and Action: Essays on the Implementation of Public Policy* (London: Methuen).
Bishop, J. and Rose J. (1992), 'Dependence, Independence and Interdependence', *Ecos*, 13(1), 14–19.
Bishop, K. and Philips, A. (1993), 'Integrating Conservation, Recreation and Agriculture Through the Market Place', *Ecos*, 14(2), 36–46.
Brenton, M. (1985), *The Voluntary Sector in British Social Services* (Harlow: Longman).
British Wildlife: see Note 2 above.
Brooke, R. (1989), *Managing the Enabling Authority* (Harlow: Longman).
Brotherton, I. (1989), 'What Voluntary Approach?', *Ecos*, 10(2), 36–40.
Brown, R. (1992) 'Site Protection in Northern Ireland: Why is Progress So Slow?', *Ecos*, 13(2), 30–5.
Buckley, P. (1989), 'Planning for Conservation in Kent', *Ecos*, 10(2), 21–6.
Burton, T. (1990), 'Sea Changes in Planning Policy', *Ecos*, 11(1), 52–3.
Butterfly Conservation *et al.*, (1993), *Biodiversity Challenge: An Agenda for Conservation in the UK* (Sandy: RSPB).
Caldwell, N. (1993), 'Cyngor Cefn Gwlad Cymru: Trying to Break the Mould', *Ecos*, 14(3/4), 42–7.
Carson, R. (1962), *Silent Spring* (London: Hamish Hamilton).
Clark, A. and O'Riordan, T. (1989), 'A Case for a Farm Conservation Support Unit', *Ecos*, 10(2), 30–5.
Colman, D., Froud, J. and O'Carroll, L. (1992), *Comparative Effectiveness of Conservation Mechanisms* (Manchester University: Dept of Agricultural Economics).
Colman, D., Froud, J. and O'Carroll, L. (1993), 'The Tiering of Conservation Policies', *Land-Use Policy*, 281–92.

Colman, D. and Lee, N. (1988), *Evaluation of the Broads Grazing Marshes Conservation Scheme* (Manchester University: Dept of Agricultural Economics).

Cox, G. and Lowe, P. (1984), (1984) 'Agricultural Corporatism and Rural Conservation', in T. Bradley and P. Lowe (eds), *Locality and Rurality* (Norwich: Geo Books).

Cox, G., Lowe, P. and Winter, M. (1986), Agriculture and Conservation in Britain: A Policy Comunity Under Seige', in G. Cox *et al.* (eds), *Agiculture: People and Policies* (London: Allen & Unwin).

Cox, G., Lowe, P. and Winter, M. (1990), (*The Voluntary Principle in Conservation* (Chichester: Packard Publishing).

DoE (1985), *Lifting the Burden* (London: HMSO) Cmnd 9571.

DoE (1990), *This Common Inheritance* (London: HMSO) Cmnd 1200.

DoE (1992), *This Common Inheritance: The Second Year Report* (London: HMSO) Cmnd 2068.

DoE (1994a), *Sustainable Development: The UK Strategy* (London: HMSO) Cmnd 2426.

DoE (1994b), *Climate Change: The UK Programme* (London: HMSO) Cmnd 2427.

DoE (1994c), *Biodiversity: The UK Action Programme* (London: HMSO) Cmnd 2428.

DoE (1994d), *Sustainable Forestry: The UK Programme* (London: HMSO) Cmnd 2429.

Dixon, J. (1992), 'Environmentally Sensitive Farming – Where Next?', *Ecos*, 13(3), 15–19.

Ebrahim, A. and Elliott, G. (1991), 'Energy Policy and Bird Species', *Ecos*, 12(4), 21–8.

Evans, D. (1992), *A History of Nature Conservation in Britain* (London: Routledge).

Farrington, J. *et al.*, (1993), *Reluctant Partners: NGOs, the State and Sustainable Agricultural Development* (London: Routledge).

Flynn, A. and Lowe, P. (1992), 'The Greening of the Tories: The Conservative Party and the Environment', in W. Rüdig (ed.), *Green Politics Two* (Edinburgh: Edinburgh University Press).

Gibbons, D. W. *et al.*, (1993), *The New Atlas of Breeding Birds in Britain and Ireland: 1988–91* (London: Poyser).

Goode, D. (1989), 'Learning from Cities: An Alternative View', *Ecos*, 10(4), 42–8.

Grove-White, R. (1992), 'Environmental Debate and Society – The Role of NGOs', *Ecos*, 13(1), 10–14.

Harvard, M. and Ferns, P. (1993), 'Cardiff Bay: A Cautionary Tale', *Ecos*, 14(2), 47–52.

Haskins, L. (1991) 'Canford Heath – Highly Publicised, Poorly Understood', *Ecos*, 12(1), 80–1.

Healey, P., Davouchi, S., O'Toole, M. Tarranoyln, S. O., Usher, P. (1992), *Rebuilding the City* (London: Spon).

Hill, J. and Jordan, A. (1993), 'The Greening of Government: Lessons from the White Paper Process', *Ecos*, 13(3/4), 3–9.

Johnston, M. (1991), 'The Forest We Live In', *Ecos*, 12(1), 65–8.

King, D. S. (1993), 'The Conservatives and Training Policy 1979-92', *Political Studies*, 41, 214–35.

Lomax, P. (1990), 'Planning for Nature Conservation in Leicester', *Ecos*, 11(4), 20–6.

Lowe, P. (1983), 'Values and Institutions in the History of British Nature Conservation', in A. Warren and F. B. Goldsmith (eds), *Conservation in Perspective* (New York: Wiley).

Lowe, P., Cox, G. MacEwen, M. O'Riordan, T. and Winter, M. (eds) (1986), *Countryside Conflicts: The Politics of Farming, Forestry, and Conservation* (London: Gower & Maurice Temple Smith).

Lowe, P. and Goyder, J. (1983), *Environmental Groups in Politics* (London: Allan & Unwin).

Lowe, P., Murdoch, J,. Marsden, T., Muntaal, N. and Flynn, A. (1993), 'Regulating the New Rural Spaces: The Uneven Development of Land', *Journal of Rural Studies*, 9, 205–222.

Maguire, F. and Barkham, J. (1991), 'The Coastal Environment: Managing Sea-Level Rise', *Ecos*, 12(2), 22–6.

Malpass, P. and Marie, A. (1990), *Housing Policy and Practice*, 3rd edn. (London: Macmillan).

Marren, P. and Rich, T. (1993), 'Back from the Brink – Conserving our Rarest Flowering Plants', *British Wildlife*, 4, 296–304.

Marsden, T., Murdoch, J., Lowe, P., Munton, R. and Flynn, A. (1993), *Constructing the Countryside*, (London: UCL Press).

Marsh, D. and Rhodes, R. A. W. (1991), *Policy Networks in British Government* (Oxford: Oxford University Press).

Mayes, B. and Smith, R. (1990), 'Conservation Carve-Up', *Ecos*, 11(2), 2–6.

McCormick, J. (1991), *British Politics and the Environment* (London: Earthscan).

Millward, A. (1990), 'Urban Horizons', *Ecos*, 11(2), 17–19.

Nicholson-Lord, D. (1987), *The Greening of Cities* (London: Routledge).

O'Riordan, T. (1989), 'Nature Conservation Under Thatcherism', *Ecos*, 10(4), 4–8.

Owens, S. (1993), 'Planning and Nature Conservation: The Role of Sustainability', *Ecos*, 13(3/4), 15–22.

Pearce, D. W. *et al.* (1989), *Blueprint for a Green Economy* (London: Earthscan).

Potter, C (1993), 'Pieces in a Jigsaw: A critique of the New Agri-Environment Measures', *ECOS*, 14(1), 52–4

Pepper, S. (1993), 'Scottish Natural Heritage: The Velvet Glove', *Ecos*, 14(3/4), 36–41.

Petersen, J. (1990), 'Heads Above Water', *Ecos*, 11(2), 49–50.

Potter, C. (1993), 'Pieces in a gigsaus: A critique of the Neww Agri-Environment Measures', *Ecos*, 14(1), 52–4

Pye-Smith, C. and Rose, C. (1984), *Crisis and Conservation: Conflict in the British Countryside*, (Harmondsworth: Penguin).

Ratcliffe, D. (1989), 'The Nature Conservancy Council 1979–1989', *Ecos* 10(4), 9–15.

Ratcliffe, D. (1993), 'Nature Conservation and Afforestation Policy', *Ecos*, 14(2), 19–22.

Rawcliffe, P. (1992a), 'Swimming with the Tide – Environmental Groups in the 1990s', *Ecos*, 13(1), 2–9.

Rawcliffe, P. (1992b), 'Lessons from the Bogs – What Now for the Peat Campaign?', *Ecos*, 13(2), 41–7.

Robinson, M. (1992), *The Greening of British Party Politics* (Manchester: Manchester University Press).

Robson, B. (1989), *Those Inner Cities* (Oxford: Oxford University Press).

Rose, J. (1990), 'Pocket Parks – Countryside Conservation by Local People', *Ecos*, 11(1), 7–11.

Rotherham, I. (1990), 'An Endangered Species? Natural History Societies and Nature Conservation', *Ecos*, 11(2), 20–5.

Rothwell, P. and Housden, S. (1990), *Turning the Tide – A Future for Estuaries* (Sandy: RSPB).

Rowell, T. A. (1991), *SSSIs: A Health Check* (London: Wildlife Link).

Rydin, Y. (1993), *The British Planning System: An Introduction* (London: Macmillan).

Rydin, Y. and Grieg, S. (1995), 'Talking Past Each Other?'. Local Environmentalists in Different Organisational Contexts, *Environmental Politics*, 4, in press.

Sands, T. (1990), 'Nature Conservation's Political Clout', *Ecos*, 11(2), 26–8.

Shirely, P. and Knightsbridge, R. (1992), 'Consultants and County Trusts – Unhappy Bedfellows?', *Eco*, 13(1), 27–31.

Smith, T. (1990), 'The County Trusts Forty Years On', *Ecos(2)*, 11, 12–16.

Smyth, B. (1987), *City Wildspace* (London: Hilary Shipman).

Stoker, G. (1991), *The Politics of Local Government*, 2nd edn. (London: Macmillan).

Stoker, G. and Young, S. C. (1993), *Cities in the 1990s* (Harlow: Longman).

Thornley, A. (1991), *Urban Planning under Thatcherism: The Challenge of the Market* (London: Routledge).

Tompkins, S. (1989), *Forestry in Crisis* (London: Christopher Helm).

Tompkins, S. (1993), 'Conifer Conspiracy', *Ecos*, 14(2), 23–7.

Turner, R. (1993), ' Britain's Forests – An Age of Enlightenment?', *Ecos*, 14(2), 14–19.

Tyldesley, D. and Collis, I. (1990), 'Nature, Conservation and Local Government – A Review of Progress', *Ecos*, 11(4), 3–20.

Tyler, S. (1992), 'Dysgu Gwers: Lessons Learned from Welsh Planning', *Ecos*, 13(2), 24–9.

Ward, S. (1993), 'Thinking Global, Acting Local? British Local Authorities and their Environmental Plans' *Environmental Politics*, 2, 453–78.

Warren, L. (1991), 'Marine Nature Reserves: Fact or Fiction', *Ecos*, 12(2), 35–9.

Wilson, J. and Fuller, R. (1992), 'Set-Aside: Potential and Management for Wildlife Conservation', *Ecos*, 13(3), 24–9.

Woodford, G. and Oliver, P. F. (1992), 'Stalling the Hereford Bypass – Turning Point or False Hope?', *Ecos*, 13(13), 19–22.

Young, S. C. (1992), *Changing Models of Government at the Local Level in Britain* (Manchester: The Politics Association Resource Centre).

Young, S. C. (1993), *The Politics of the Environment* (Manchester: Baseline Books).

Young, S. C. (1994a), 'The Environment', in P. Allan *et al.*, *Focus on Britain 1994* (Deddington: P. Allan).

Young, S. C. (1994b), 'An Agenda 21 Strategy for the UK?', *Environmental Politics*, 3, 325–34.

Yull, L. (1993), 'Forestry Policy and Practice', *Ecos*, 14(2), 27–31.

15 The Tragedy of the Common Fisheries Policy: UK Fisheries Policy in the 1990s

Anthony Stenson and Tim S. Gray

INTRODUCTION

The notion of the Tragedy of the Commons is familiar to those engaged in the study of environmental politics. Put simply, the concept, originated by Garret Hardin in a seminal 1968 article in the US journal *Science*, states that in a commons (a resource which no one owns but all have the right to use for their own benefit), the inevitable result of each person pursuing their own self-interest is the destruction of the commons for all. Hardin suggested two solutions: 'We might sell [the commons] off as private property. We might keep them as public property, but allocate the right to use them' (Hardin, 1968: 1245). As Wijkman wrote (1982: 524), some commons, by their very nature, do not allow the first course of action to be taken. Sea fisheries are just such an example. Here, the only option is to allocate the right to prosecute stocks. This is the system in operation in almost all the world's fisheries today, including the fisheries surrounding the UK.

The significant fact about fisheries is that they are usually an international commons, and, by and large, do not come under the jurisdiction of single sovereign governments. That is, fisheries tend to be spread over more than one jurisdiction. This makes agreements on how to protect the commons harder to come by, since they necessitate some kind of restriction on national sovereignty.

This chapter is about the current fisheries conservation policy of the UK and the politics surrounding it. However, it should be noted that when we use the word 'conservation', we do not use it in the same sense used when we speak about the conservation of an endangered species. None of the fish species around the UK is in danger of extinction. Moreover, conservation in relation to UK fisheries is unambiguously human-instrumental. Fish are conserved not in order to maintain biodiversty, or because fish are deemed to have innate value, but only to enable humans to optimalise catches in the long term. Paradoxically, fish are conserved so more of them can be killed.

The aim of the EU's CFP is to ensure the sustainable development of the fishing industry. However, in the UK, as elsewhere, this aim has not been achieved. Critics point to the total lack of an effective long term strategy, for which both the EU and the UK Government can be blamed, which has left the industry in a permanent state of crisis. By common consent, stocks are overexploited and there is too much capacity in the fleet. Markets are unstable, partly as a result of some of the tools being employed as part of the conservation regime. Fisheries conservation has degenerated into a state of permanent crisis management, and the industry has lost all confidence in those who regulate it.

This chapter is set out as follows: the following section deals with the CFP, its history and mechanisms; the next deals with the failures of the CFP, with specific regard to the UK; then we discuss the way the UK implements the CFP, which is followed by an assessment of the near future for the CFP and the UK's fisheries sector; and we finish with a short conclusion.

THE CFP

Since 1972 Britain's fisheries policy has been set in the framework of EU fisheries policy. A knowledge of the CFP is crucial, therefore, to an understanding of the policy of the UK. In this section we set out the working of the CFP, with specific regard to the UK, and begin with a short history.

History of the CFP

The fisheries policies pursued by the individual member states have always been based on a very weak 'lowest common denominator' policy at the centre, due to a lack of common interest among the various nations' fishing sectors. This has led to incoherence and confusion, with different measures being implemented in different member states.

The CFP's history goes back to 1966 when the Commission presented proposals to the Council for a common fisheries policy. The Treaty of Rome provides the rather ambiguous basis for the CFP, which is the same basis as that for the CAP. In Article 38(1) it is signified that all articles pertaining to agriculture also pertain to fisheries. Therefore, Article 3(d), requiring the adoption of a common policy in the sphere of agriculture, also requires a common fisheries policy (Leigh, 1983; Elles and Farnell, 1984; Wise, 1984).

Differences between the original six members of the community on the issue and the fact that fisheries were not a high priority for them meant that it was four years before the first regulations on fisheries were passed. These first regulations were hurriedly passed on 30 June 1970, the day that access negotiations began for Denmark, Ireland, Norway and the UK. There was a significant political reason for this timing. All countries entering the community must accept the *acquis communautaire* (the sum of the community's achievements and area of competence). Candidate states are entitled to influence the *acquis* as soon as negotiations begin (Elles and Farnell, 1984; Churchill, 1987). The four – all with significant fishing interests – would have preferred to keep control of fisheries under national jurisdiction. The passing of the regulations at this time meant that they had no choice: they had to accept the EC's powers in relation to fisheries or refuse to join. The UK, Ireland and Denmark chose the former, and Norway the latter, option.

The most significant regulation was 2141/70, the structural regulation governing the catching side of the industry. This regulation both affirmed the principle of equal access, and handed over control of conservation measures to the EC. The equal access principle was justified by Article 7 of the Treaty of Rome, which prohibits discrimination on the ground of nationality, and it means that all the waters of member states are open to all EU fishermen. This principle was unacceptable to the applicant states in 1970, and a compromise was reached whereby the 1972 Treaty of Accession made a number of concessions, known as derogations. Article 100, for example, permitted member states to restrict fishing in a six-mile coastal zone to boats fishing traditionally in those areas from local ports. Article 101 provided for the extension of that zone to 12 miles in certain areas which were considered to be heavily dependent upon fishing. This included much of the UK. These derogations, a good example of the 'progress by lowest common denominator' model of common policy formation, whereby concessions are made to make the policy more amenable to all parties, were to last until the end of 1982. Article 102 called for the council to 'determine the conditions for fishing with a view to ensuring protection of the fishing grounds and conservation of the biological resources of the sea'. This was to be done by 'the sixth year after accession at the latest'(i.e., 1978). Although ambiguously phrased, this article was taken as a committal by the member states to institute a fully comprehensive CFP by 1978 (Leigh, 1983; Elles and Farnell 1984, Wise, 1984; Churchill, 1987; Holden, 1994).

Agreement on a complete CFP was not, however, reached until 25 January 1983. It is probable that the impasse would have dragged on longer, had it not been for the fact that the derogations that the three won in the access negotiations a decade before expired on 31 December 1982. Thus there was an urgent political need to settle.

The Regime

The regime agreed upon in 1983 is basically the same as the one we have today. The main components of the 1983 package were a system of total allowable catches (TACs) allocated for each species of economic significance each year and divided up between the member states according to a set formula; technical conservation measures; multi-annual guidance programmes (MAGPs) to reduce capacity; and Regulation 2057/82 on enforcement. We will discuss technical measures and enforcement later; let us now turn to TACs and MAGPs.

TACs

TACs are set annually by the Council of Ministers and the Commission on the basis of scientific advice. Up until a few years go, the scientists' recommended TACs tended to be revised upwards by the Council and Commission in the interests of political expediency. However, this tendency has decreased in recent years as the Council and Commission have attempted to pursue a more 'scientific' policy (Symes, 1991). The result of this has been dramatically reduced TACs since 1989, and a subsequent upheaval within the industry. A few TACs are 'analytical TACs', based on detailed empirical data on fish stocks; most are what are known as 'precautionary TACs', based mainly on historical catch data. Full data is not collected for these TACs because it is not considered worth the (quite formidable) expense. A 'scientific' policy is therefore being pursued on the basis of inadequate scientific information. As Wijkman observed in 1982, 'when the size of the resource is unknown, it may in practice be difficult to convince those who exploit the resource of the need to limit their harvest' (Wijkman, 1982; 517). Such a situation is now developing in the UK at the moment; fishermen, convinced that the seas are teeming with fish, are irate at what they perceive to be overconservative stock estimates by the scientists.

In 1983 TACs were divided up into national quotas. It was agreed that the carve-up would be permanent and each country would receive the same percentage of each species every year. This is what is known as the principle of 'relative stability', and the agreement was made in order to ensure continuity within the industry, and to prevent the protracted wrangling any renegotiation would inevitably involve.

The carve-up was done on the basis of three factors: historic catches, of which the reference period would be 1973–78; special preferences for those areas whose economies were heavily dependent upon fisheries (Hague preferences); and some compensation for jurisdictional losses, meaning compensation in fishing opportunities for losses sustained as a result of the 1977 global move to 200-mile fishing limits. For the UK the Hague preference areas lie north of a line running from Bridlington in Yorkshire to Morecambe in Lancashire. This means that these areas are allocated a higher proportion of stocks than other areas, in times of severe cutbacks in TACs, as they are deemed to be more heavily dependent on fishing. When the TAC has been divided up into national quotas, each country then allocates their quota as they see fit.

MAGPs

The conventional wisdom sees the greatest threat facing the EU and UK fishing industries as that of overcapacity. The purpose of MAGPs is to trim that overcapacity and bring fleet sizes more into line with fishing opportunities. Growth in capacity must therefore be halted and reversed. The first MAGP was for the period 1983–6, and the second for the period 1987–91. Both failed in EU-wide terms and in UK terms, the UK failing to meet its target for the second programme by 11 per cent. The government has been criticised from many quarters for this outcome, as it failed to stump up sufficient funds for an effective programme of vessel decommissioning. It has been said that an effective decommissioning programme a few years ago would have avoided many of the problems faced by the industry today.

The 1993–6 MAGP demands a reduction in capacity in the UK fleet of 19 per cent overall compared with 1992. The new MAGP allows up to 45 per cent of the reduction to be achieved through measures other than the removal of vessels from the fleet. This is an improvement as, under the old system, MAGP targets could be achieved simply by decommissioning vessels which were in any case old and rarely used; fishing effort was, therefore, unaffected. Now action can be taken under MAGP guidelines directly on the extent to which capacity is put to use. This can include such measures as days-at-sea limitation (House of Commons Select Committee on Agriculture, 1993).

The 1992 CFP

Regulation 170/83 stated that there should be a mid-term review of the CFP by the end of 1992. Negotiations lasted for one year before the council agreed a new Regulation (3769/92) in December 1992, in which the biggest change was the greater emphasis placed on effort reduction through MAGP programmes. In addition, new mechanisms were provided for the development of more long term strategies on TACs and quotas. In the next few years we may see the appearance of multi-annual TACs and multi-species TACs. The final significant development of the mid-term review was a commitment to strengthen enforcement across the EU.

THE FAILURE OF THE CFP

The official objectives of the CFP, as they appear in Article 1 of Regulation 101/76, which is a revised version of the original structural regulation 2414/70, are vague: 'to promote harmonious and balanced development of this industry within the general economy and to encourage rational use of the biological resources of the sea and of inland waters'. The 1983 settlement was more specific: 'to ensure the protection of fishing grounds, the conservation of the biological resources of the sea and their balanced exploitation on a lasting basis and in appropriate economic and social conditions' (Regulation 170/83). The 1992 Regulation is yet more explicit in its aims, perhaps indicating that the CFP is moving beyond the lowest common denominator stage and growing in effectiveness and purpose: 'the objective should be to provide for rational and responsible exploitation of living aquatic resources and of acquaculture, while recognizing the interest of the fisheries sector in its long-term development and its economic and social conditions and the interest of consumers taking account of the biological constraints with due respect for the marine ecosystem' (Regulation 3760/92). However, it is too early to assess whether these latest aims have been met. For this reason, we shall confine ourselves to an evaluation of the CFP based on the 1983 objectives.

The 1983 aims are basically two-fold: to conserve fish in the interests of optimalising catches in the long term and to ensure the long term viability of the industry. The CFP has failed to fulfil these aims. By common consent, EU fish stocks are overexploited and in an almost permanent state of crisis; there is too much capacity in the EU fleet overall; and the fishing industry, including the UK's, is in a hazardous position, with its long term future in doubt. The problems of the CFP have been legion, and we shall now examine them with specific regard to the UK fishing industry.

TACs, Quotas, Discards and Blackfish

The problem of discards, along with the growing problem of blackfish (black market fish), is caused mainly by the system of TACs and quotas and its accompanying ban on the landing of overquota and undersized fish. Discards are a particular problem in the UK because the waters surrounding the country are generally what are called 'mixed fisheries' (i.e., they contain fish of many different species). This means that it is easy accidentally to catch fish other than the target species. Often, this 'by-catch' is of fish for which there is no remaining quota, and fishermen in such a situation have to throw the fish back into the sea. Once fish have been ensnared in a fishing net, they have a virtually nil chance of survival. Discards therefore result in hundreds of thousands of dead fish being dumped in the sea (Agriculture Committee, 1993), which represents a colossal waste of good fish.

The blackfish market (i.e., surplus fish which would otherwise be discarded but is sold in unauthorised deals), by all accounts ballooned in the last few months of 1993, and reached the stage where prices on the legitimate market and the enforcement of conservation regulations were being undermined. The press carried reports of a huge blackfish operation involving all sectors of the industry from fishermen to fishmonger, and a Scottish MEP called for a government inquiry into the situation. A large market in black market fish is obviously deleterious to efforts at conserving stocks. Blackfish landings are not recorded and therefore undermine attempts to measure stocks, and they make a mockery of the elaborate systems of control employed to control fishing in the mainstream market.

The Norwegians have instituted a ban on discards in their waters, but UK and EU policy is against such a move: firstly because it would be difficult to enforce; secondly because if in order to avoid increasing the blackfish market, it entailed setting up alternative marketing arrangements for overquota fish, this would mean that fishermen would get a reduced price for such fish and would therefore be tempted to discard anyway; and thirdly, if overquota fish were permitted on to the mainstream market, the whole system of quotas would be made meaningless, as fishermen would have no incentive not to catch species for which they had no remaining quota (Agriculture Committee, 1993). Additionally, it would deflate the market price of quota fish by adding to the supply.

In the opinion of the Agriculture Committee, TACs have consistently been set at levels which have maintained stock overexploitation: 'In general the Council has not fixed TACs which correspond to reductions in the rate of fishing on the stocks, but has preferred to fix them at levels which maintain employment, even though this results in the stocks remaining over-exploited' (Agriculture Committee, 1993, xx). The Committee also reported, as did many of the inquiry's witnesses, that TACs are inappropriate for the mixed demersal (sea-bed) fisheries typical of UK waters, where by-catches form a high proportion of overall catches. For this reason, although the Committee came out in favour of a system of TACs and quotas, it also said that they will never provide an effective method of fisheries management and stock conservation.

It seems that TACs and quotas are being used as a method of fisheries management across the EU because they are basically the only method whereby stocks can be shared out on a settled basis between the member states. The motive behind this is to ensure stability and continuity within and between the member states' fishing industries, and not, primarily, to conserve fish. Holden writes that the regime of TACs and quotas adopted by the 1983 CFP was the result of a debate not about how to manage the resource effectively but about how to deal with the political problems caused by two developments during the 1970s: firstly, the provisions on equal access which were instituted just before Denmark, Ireland and the UK joined the EU and, secondly, the global move to 200-mile fishery limits. The first factor created a 'Community Sea', which led UK, Danish and Irish fishermen to feel that they had been cheated out of the benefits they would have gained from the second. A system of TACs and quotas would assuage their fears and guarantee them a certain amount of fish per year. However, the legal fact of equal right of establishment across the EU, backed up by the 'Factortame' lawsuit (discussed more fully below), has rendered TACs and quotas futile; boats from any EU country can legally register in any other, and take advantage of that country's quota (Holden, 1994: 258–9).

Technical Measures

EU legislation involves the employment of certain technical measures aimed at conservation. Legislation specifies minimum mesh sizes (MMSs) for each fisheries area and species, though only for methods of fishing (trawls, Danish seines and other towed gear) which are the most significant. The idea behind MMSs is to facilitate the escape of juvenile fish from the net.

MMSs are backed up by minimum landing sizes (MLSs), which prevent the landing and sale of juveniles on shore. MLSs are popular within the industry, and are relatively cheap and easy to enforce, as inspection takes place on land. However, it seems that across the EU both MMSs and MLSs are too low, and ought to be raised in order to enable fish to reach breeding age, and breed at least once, before they come under threat of capture. The

British industry is largely in favour of such a measure. Gear specialist John Ashworth told us that only a small increase in sizes would be needed. Some feel that the reason the EU has not raised MMSs and MLSs has been to appease the Spanish and Portuguese, whose fishermen are believed to deliberately target juveniles to satisfy the tastes of their customers.

Technical measures are looked upon with a great deal of enthusiasm in the UK, where many fishermen feel that gear regulations can almost single-handedly achieve effective conservation. Even UK scientists, who do not generally hold this opinion, have estimated that an increase in mesh sizes to 120mm for cod stocks could lead to an increase in stocks of 25 per cent in a few years (Symes, 1991). In their December 1993 proposals to the government, the National Federation of Fishermens' Organisations (NFFO) relied heavily on proposals for gear regulation.

Enforcement

Enforcement represents the most disappointing aspect of the CFP. Implementation of fisheries regulations is left to the individual member state, but enforcement is extremely uneven across the EU. The Commission Report on enforcement in March 1992 criticised all the member states' records, but stated that the UK had the most stringent of controls, complimenting them on their 'vigour and efficiency' (EC [92] 394 final, quoted in Agriculture Committee, 1993). This vigour is a source of a great deal of resentment among UK fishermen, who consider themselves to be at a disadvantage when compared with other countries whose enforcement standards are very much below UK standards. France, and particularly Spain, are thought to be the major culprits. Spain has only 17 inspection officials, all based in Madrid, compared with the UK's 170 officials based in fishing ports. The Agriculture Committee made trips to France and Spain in order to gather evidence on enforcement, and found both grossly inadequate enforcement mechanisms and much evidence of shady practices among Spanish fishermen: 'we attach credence to most, if not perhaps all, of the evidence we received about malpractice on the part of a number of Spanish vessels. Anecdotal evidence from UK fishermen, and from UK and French fisheries protection authorities, is borne out by the statistics' (1993, xxxvi).

UK CONSERVATION POLICY

The UK is subject to the CFP regime but, like all other member states, it is responsible for the implementation of the CFP and, as with other common policies, it is permitted to operate its own conservation measures over and above the CFP. There are four basic components to the UK's implementation of the CFP.

Enforcement

The first component is the UK enforcement structure, widely regarded as the most zealous and effective in the EU. Enforcement takes place on land and at sea. On land, inspectors carry out random inspections of vessels, checking on minimum landing sizes and some technical measures and ensuring that landings are accurately reported. In English and Welsh waters, Royal Navy fishery protection vessels have powers to board ships and check that MMSs and various other technical measures are being complied with. Vessels also make sure that there are no boats fishing where they ought not to be. Within a three-mile coastal zone, local Sea Fisheries Committees take over from the Navy. In Scottish waters, the same duties are carried out by civilian patrols. Trials are being carried out throughout the EU on the use of satellites as a means of spotting vessels which are fishing in areas where they are prohibited (Agriculture Committee, 1993).

Licensing

The second component in UK implementation is the licensing system it operates. The licensing system underwent changes in 1993 as the government sought to use it as a method of reducing capacity in line with MAGP targets. A vessel fishing for a pressure stock – one which is heavily prosecuted and therefore subject to the TAC regime – must have a licence. Licences, which are allocated free of charge but can be traded and amalgamated, simply allow the bearer to fish for a certain stock, and do not specify the amount which can be caught. Licences are issued only to those boats which have a track record fishing for a particular stock, and are therefore aimed at preventing boats from starting to fish for a new species. From 1 May 1993, the licensing system was extended to boats of less than ten metres, a move welcomed by fishermens' organisations and the Agriculture Committee (1993: xlii).

Licensing is thought by some to hold the answer to the problem of fisheries management. Holden believes that the fundamental problem is one of access, and that if access to the resource is cut, stocks will recover and catches and profits will increase:

There is one objective which all fishermen want, which is to have a reasonable standard of living from operating in a profitable industry ... Fishery science shows the means by which it can be realised, which is primarily to reduce the amount of fishing by limiting the number of fishing vessels. This should be achieved by a system of limited licensing. (Holden, 1994: 261)

Decommissioning

A decommissioning scheme forms the third strand of the UK Government's policy. The CFP has long recognised the fact that the fleet must stop growing, and even be cut; yet the way this end is being pursued in the UK is far from satisfactory from the point of view of the UK fishing industry.

The UK scheme of 1993–6 provides for £25 million over three years in decommissioning grants. This is intended to bring about a reduction in capacity of 5 per cent, fulfilling about a quarter of the UK's MAGP responsibilities (the rest is to come from days-at-sea restrictions and licensing). However, the scheme is far from satisfactory. The amount made available has rightly been criticised as insufficient, especially as it is to be spread over three years. The NFFO has calculated that the Treasury will eventually recoup all the £25 million and a further £9 million in taxation from the scheme, which has not been officially challenged.

This is a familiar story to many of those involved in the fishing industry. The government has never, it seems, taken its MAGP responsibilities seriously. The first MAGP was all but ignored, and grants awarded to owners buying or modernising boats meant that the incentives to stay in the industry far outweighed the incentives to leave. Many regard the government's failure to cut, or even stabilise, capacity in the 1980s as responsible for a great deal of the troubles facing the industry today. Scottish Fishermens' Federation (SFF) leader Cecil Finn has said: 'the problems [of the industry] have arisen from the government's negligence in ignoring its EU-directed MAGP obligations in the 1980s' (*Fishing News,* 8 October 1993).

Days-at-Sea

The final component is the highly controversial, and now suspended, days-at-sea scheme, introduced under the Sea Fish (Conservation) Act 1992, which gave the minister the power to restrict days at sea for all vessels over ten metres in length. The government's intention was to freeze time at sea at 1991 levels, and then to reduce it by as much as 8.5 per cent, which is the portion of the UK's MAGP targets to be attained through effort control. A restricted days-at-sea scheme has been in operation since 1990 in the North Sea, and this resulted in tie-ups being brought in in 1991 and 1992. The EU had planned to institute a more comprehensive days-at-sea scheme for the North Sea, but agreed not to as long as the UK Government came up with its own scheme. The EU's scheme would have been a measure intended to ease short term pressure on stocks, but the UK Government intended their scheme as a structural measure, and applied it to the entire UK fleet.

In the past, and indeed now, the fishing industry has been characterised by its divisions and rivalries, but all were agreed on the evils of the Sea Fish (Conservation) Act. Fishermen felt that it would bankrupt many of them and destroy their fragile industry. However, the project was dogged from the start, and it has now been suspended pending legal clarification.

The administration surrounding the scheme, for example, faced formidable difficulties. Every vessel over ten metres had to be allocated a number of half days based on its track records (time spent fishing) in 1991. This meant collecting thousands of track records, many of which were threadbare in detail and unreliable as evidence. Moreover, 84 per cent of vessel owners appealed against their allocation, and this vastly increased the administration involved. Compounding this problem was the huge amount of protest against the Act from the industry. This protest was often striking in its depth of feeling; the methods employed by the fishermen ranged from the blockading of harbours to legal action, and it seems fair to say that the government were taken aback by the intensity of opposition to the scheme.

It was the NFFO's legal action that has done the most damage to the days-at-sea scheme. Arguing that the scheme violated the non-discriminatory norm of the EU, the NFFO managed in December 1993 to get the case referred by the High Court to the European Court of Justice (ECJ). The judgement was only a referral but, given that the case will take two years to reach the ECJ, and that the Act has been suspended by the High Court until that time, it represents a significant victory for the NFFO and its followers.

The days-at-sea scheme was intended to run alongside the existing system of TACs and quotas. Therefore, in theory, during their fewer days at sea fishermen would be able to catch as many fish as before. This is what the government claimed, arguing that the scheme would reduce effort by making fishing more efficient since fishermen would use their limited time more productively, and by-catches and discards would therefore be reduced.

So why are fishermen so implacably opposed to the days-at-sea scheme? During the campaign, they made a number of claims against the regime. Firstly, they claimed that the scheme would not serve its stated purpose of conserving fish. The NFFO asserted that 'it does directly or effectively address stock depletion problems faced by some important commercial species' (NFFO, evidence to Agriculture Committee, 1993). Fishermen, according to this view, will simply fish more intensively in the time available, search for the nearest stocks rather than for concentrations of adult fish, and surreptitiously reduce gear selectivity, thus increasing by-catches and the capture of juveniles. Secondly, not all fishermen were in a position to produce track records for 1991. This applies especially to owners prosecuting non-pressure stocks, since they were not required in 1991 to keep a record of their catches. The government said that it would respect unorthodox records in arbitrating on appeals, but this merely worsened the administrative burden of the appeals procedure. Thirdly, fishermen claimed that the Act was dangerous in that skippers would want to use up all of their entitlement and would therefore put to sea in adverse weather conditions, which would previously have kept them on dry land. Fourthly, the bad feeling which already exists between UK and foreign fishermen would increase as a result of the Act, as non-UK boats would be able to fish for their quota at times of the year when UK boats were tied up. Moreover, foreign boats would also be building up large track records which would in future be used as the basis for allocating quotas when these stocks came under the quota regime. Finally, as stated above, the government intends eventually to begin discriminating in the allocation of days-at-sea between the various sectors of the fleet, to take account of differing pressures on the various stocks, and this is a divisive policy.

For these reasons the Act has been widely regarded as a deliberate and savage attack on the UK fishing industry, intended to drastically cut capacity without having to pay out compensation. The Agriculture Select Committee's Report said: 'Our view is that days-at-sea restrictions applied to the whole fleet over ten metres in length are unnecessarily draconian, and amount to little more than decommissioning on the cheap' (1993: xlii). On 7 July 1993, Michael Jack announced the postponement until January 1994 of the days-at-sea scheme, and invited the industry to come up with its own proposals for tackling conservation. This they did but, as we indicated earlier, the NFFO also challenged the scheme in court and won a referral to the ECJ. As the scheme was intended partly to fulfil the 1993–96 MAGP obligations, and given that it may take two years before the ECJ hears the case, the scheme is all but dead.

THE FUTURE OF THE CFP AND UK FISHERIES POLICY

The CFP arrived at in 1983 is a set of derogations from the Treaty of Accession, and is officially scheduled to last until 2002. There are a number of issues which are already causing problems, however. We outline these questions below.

Norwegian Rejection of EU Membership

Norway has now rejected EU membership twice – in the referendums of 1970 and 1994. In each case, one of the main reasons for the 'No' vote was the perceived disadvantage to Norway's fishing industry of joining. The prospect of Norwegian accession posed two significant problems for the UK. First, it would have led to much greater catching opportunities in the North sea by Norwegian fishermen. Second, in order to obtain the necessary Spanish and Portuguese support for the deal with Norway, EU ministers had to agree to full integration of Spain and Portugal into the North Sea, and this posed a serious threat to the UK industry.

However, Norway's decision to remain outside the EU has caused considerable dismay to the UK pelagic sector, since Norwegian accession would have allowed British vessels to catch a significant proportion of their mackeral TAC in the Norwegian sector. At present, the UK pelagic fleet is unable to catch its full mackerel allocation because for much of the time the bulk of the fish lie out of reach in the Norwegian sector. Moreover, the Norwegian negotiating stance with the EU has been hardened by the 'No' vote, and the UK can expect both lower TACs on whitefish and mackerel, and even tougher enforcement measures by Norway of her fishing regulations.

Spanish and Portuguese Accession to EU Waters

British fishermen see their Spanish and Portuguese counterparts as lawless, with no regard for conservation. The historical evidence backs this up. Spanish and Portuguese fishermen were widely blamed for the closure of the Grand Banks, over 200 miles from the coast of Canada and the USA, necessitated by the gross overfishing and resultant collapse of these areas. They were also banned from fishing off Namibia three years ago after these grounds had been cleaned out. The problem seems to be that in the southern European countries, it is a tradition that each dinner plate must have an individual fish (as contrasted to the UK where large fish tend to be cut into steaks). Since small fish are regarded as a delicacy, and driven by market demand, the fishermen of such countries therefore target juvenile fish. This means that the fish are never allowed to reach adult age, so there are fewer and fewer recruits into the spawning biomass, and stocks collapse.

Up until 1996, Spanish and Portuguese fishermen have very restricted access to the waters of the pre-enlargement EU. Just 300 named boats (Spanish vessels) are allowed to fish in these waters, and only 145 at a time. In addition, Spanish and Portuguese vessels are banned from fishing in the "Irish Box" (i.e., the waters around Ireland including the Irish Sea).

The accession of Spain and Portugal to the EU in 1985 provided for these regulations to run until 2002, at which time new conditions would come into force. However, Spain has successfully lobbied to bring this date forward to 1996, when they will be fully integrated into the CFP and will have the right of equal access to all EU waters. Despite the reassurances from UK ministers that relative stability will be maintained and fishing effort will not be increased, this represents the arrival of the 'worst case' scenario for UK fishermen. It seems a matter of simple logic that if more boats are to be allowed in (and Spain's fishing fleet alone is half the entire EU fleet) while overall effort is to remain at the same level, then effort must be cut for all. It is hard to envision an outcome where the UK will be anything but a loser in such a situation.

Quota Hopping

Perhaps a greater threat to the present form of the CFP is the practice of 'quota hopping'. This is the practice of registering boats from one country in another, in order to take advantage of that country's quota. It is most common in the UK, where a number of Spanish-owned boats are registered. The significance of the quota-hoppers is that they weaken the principle of relative stability which underlies national quotas. If a boat can register in any EU country, national quotas are side-stepped. The government tried to deal with the problem in 1988, by stipulating in the Merchant Shipping Act that 75 per cent of the ownership of any fishing boat registering in the UK must be British. However, in the famous 'Factortame' case, the European Court ruled that this provision was contrary to EU law, and therefore null and void.

Relative Stability

An opinion now shared by many is that TACs and quotas can no longer be effective as a conservation tool. The shortcomings of the system have already been dealt with, but one point still needs to be made. This is that conservation is not the sole aim of such a system (Symes, 1991; Holden,

1994). TACs and quotas mainly serve relative stability; indeed without them, it would be impossible to maintain this principle. Relative stability is basically a carve-up of the TAC between the member states which remains roughly equal each year. The idea is to share out the catch between member states on a stable basis. It is a principle driven by political considerations: it guarantees a certain amount of fish per country per year, over which that country exercises control (without this, control of all stocks would be exercised by Brussels). TACs and quotas have undoubtedly failed as a method of stock control and conservation, yet it seems that they are the only means of ensuring relative stability. The choice, then, seems to be effective conservation or relative stability (weakened as it is by quota hopping), and at the moment, relative stability is the favoured option: the triumph of politics over sustainability.

An Industry Divided

Due to differences in sea conditions, the UK's fishing grounds are extremely varied, and as a result divisions within the industry are perhaps inevitable. Even so, the catching side of the industry is split to a surprising extent, considering that the political difficulties facing the industry are broadly similar. Part of the answer lies in the fact that there is no UK-wide organisation to look after fishermens' interests. Instead, there is a plethora of local federations which are affiliated either to the SFF, or to the NFFO (in England and Wales). These two organisations are not trade unions, at least in the usual sense, and they are poorly coordinated, most of the time treating each other as competitors. In addition, local organisations and sector organisations, such as the Scottish White Fish Producers' Association (SWFPA), operate with a great deal of autonomy. The SWFPA, although affiliated to the SFF, is almost its rival, and often pursues a separate policy line.

Despite the fact that the whole fishing industry despises the Sea Fish (Conservation) Act, therefore, there has been little united industry action over it. While the NFFO concentrated on a legal challenge, the SFF chose to run a publicity campaign alerting the UK public to the destruction, as they see it, of the historic British fleet. The SFF was decidedly lukewarm over the NFFO's legal action, and did not contribute to the NFFO 'Fighting Fund'. Both these organisations, however, initially campaigned against the UK Government's implementation of the CFP rather than the CFP itself. By contrast, the SWFPA sees the problem as a European one, and is supporting the Save Britain's Fish (SBF) campaign, which is pushing for British withdrawal from the CFP and the repatriation of the UK's fishing grounds.

Withdrawal from the CFP?

It is the last strategy which finds most favour with many UK fishermen. Although there is a great deal of hostility towards the UK Government, the effects of the CFP upon the UK's fishing industry have brought out a great sense of grievance against the EU. The EU is seen as the crux of the problem, and the SBF campaign, backed by the SWFPA and now by the NFFO, is turning these feelings into a political battle. The campaign sees a grand design on the part of the Commission to destroy the UK's fishing industry in order to make room for the Spanish in the North Sea and other areas. The campaign regards overcapacity as a myth, and claims that the North Sea and most of the UK's fishing grounds are not, as the scientists say, overexploited, but are in fact full of fish. It is a mixture of faulty research and political pressure which accounts for the discrepancy.

The answer, according to the SBF campaign, the SWFPA and the NFFO, is to leave the CFP, reclaim the UK's fishing grounds and institute our own conservation policies. The fact that the major breeding grounds of nearly all North Sea stocks lie in UK waters could then be used by the UK to dictate conservation measures to other countries.

This radical strategy is unlikely to succeed. It may be desirable for the UK to leave the CFP, since it has done little good for the UK's fishing industry, failing to conserve stocks and leaving the industry in a worse state than when the policy began. However, leaving the CFP would necessitate the setting-up of an alternative regime through international negotiation. This would be fraught with difficulties. There is no reason to believe, as Holden points out, that a national regime would be any more effective (Holden, 1994: 235–8). Moreover, to leave the CFP, and more importantly to abrogate the principle of equal access, would be contrary to the Treaty of Rome and the whole set of principles underlying the EU, such as non-discrimination between the citizens of member states and an 'ever-closer union'. Therefore, to leave the CFP, despite the protestations to the contrary of the SBF, the UK would almost certainly have to leave the EU. It hardly needs to be said that this is unrealistic.

The campaign's belief in a grand design on the part of the EU to destroy the UK's fishing industry in favour of Spain's is equally controversial, yet not entirely without foundation. The evidence advanced by campaign members is three-fold. Firstly, they point to high levels of unemployment in Spain; giving Spain increased access to fishing opportunities would reduce unemployment in coastal areas. Secondly, they rightly say that Spain has a large and influential fishing lobby within the EU. Thirdly, they also correctly point out that the Spanish are at present undertaking a large programme of modernisation of their fleet, with help from EU funds. The scale of this programme suggested to many, even before the agreements made in 1994, that the Spanish at least believed that the day when they could fish freely in the North Sea was not far off.

The existence of the SBF campaign and its support from the SWFPA is regarded as regrettable by some, as it represents a deep division in the industry over the causes of the industry's problems and the way to go about solving them. Given the likelihood that EU control of fisheries management is here to stay, most of those at the head of the industy's representative bodies, such as the Shetland Fishermens' Association's John Goodlad, feel that the best way forward is to stay in the CFP and change it from within, improving the interpretation and implementation of the CFP by the British Government. Ordinary fishermen, however, would probably rejoice at the secession of the UK from the CFP.

The SBF campaign is significant because it represents a profound anti-European feeling among fishermen which has grown considerably in the last few years. The letters page of *Fishing News* is full of denunciations of the sharp practices of foreign fishermen, and righteous indignation at the underhand methods Brussels is using to destroy the UK fleet.

The Future

The future for the UK fishing industry does not look bright. The next few years will see developments which hold little but gloom for fishermen and, in such a climate of almost continuous crisis, investment in this industry is likely to remain depressed, since outside investors are unwilling to put capital into an industry with so uncertain a future, and those already involved in the industry are also understandably reluctant to spend money on boats, equipment or modernisation. The result of this will be a fleet which will continue to age rapidly, while certain other countries, notably Spain, will be creating a modern, efficient fleet which will give them a great advantage over the UK in years to come.

CONCLUSION

The present system for the conservation of the fisheries around the UK is, in one sense, working. The commons is not, at present, threatened with destruction and the potential tragedy has not occurred yet. But the makings of a tragedy, identified by Hardin in 1968, are present, and the CFP can be identified as a failure for two reasons: firstly, although stocks are not in imminent danger of being wiped out, they are well below optimal levels, indicating bad management; secondly, there have been severe economic and social costs. In the UK, for example, draconian policies mean that whole fleets are threatened with decimation, along with their communities.

In our view, the failure of fisheries policy in the UK is in large part due to the lack of common interest on the part of the member states. The CFP remains a feeble 'lowest common denominator' policy, there being a lack of political will to make anything greater out of it. The result of this has been that UK fishermen have had to suffer zealous enforcement of insufficient rules while those of other countries have been able to all but ignore the rules and get away with it.

Given the political problems at the European level, perhaps the CFP's failure has been inevitable. Nevertheless, the British Government cannot escape some of the blame for the policies they have (or have not) brought into being under the umbrella of the CFP. The fact that the CFP is little more than a set of guidelines means that individual national governments must take responsibility for the policies they impose under it. Although the UK Government's enforcement of the regulations has been vigorous, the policies being enforced have not been well chosen. A good example is the way in which the government has consistently failed to achieve its MAGP obligations. The level of these obligations are decided at EU level, but national governments are charged with the responsibility of implementing policies to achieve the targets. The UK Government did not try to achieve these targets in the 1980s, and is only offering a derisory decommissioning package now.

If the government's MAGP record is poor, the dreaded days-at-sea scheme promises to be a disaster. If ever implemented, it will at best achieve only minimal conservation benefits, and this only at the cost of decimating the fleet in an arbitrary way.

The management of fish stocks in UK waters could, in our view, be better achieved if the member states of the EU took back control of stocks in their national exclusive economic zones. The problems involved in common policy formation and implementation would be avoided, and a more effective policy could be more easily organised, this time on a national and perhaps even regional level. Bilateral agreements between member states could be used to allow a certain amount of inter-fishing in EU waters.

Given wider developments in European integration, however, it seems clear that this will not happen. The EU will not surrender its competence in a sector it has controlled for over 15 years. Perhaps the CFP will gain in strength as the EU-wide fishing industry comes closer together and forges some common interest. Such a development is, however, not easily foreseeable.

References

Churchill, R. R. (1987), *EEC Fisheries Law* (Dordrecht: Martinus Nijhoff).

Elles, J. and Farnell, J. (1984), *In Search of a Common Fisheries Policy* (Aldershot: Gower).

Hardin, G. (1968), 'The Tragedy of the Commons', *Science*, 162, 1243–8.

Holden, M. (1994), *The Common Fisheries Policy: Origin, Evaluation and Future* (Oxford: Fishing News Books).

House of Commons Select Committee on Agriculture (1993), *Sixth Report: The Effects of Conservation Measures on the UK Sea Fishing Industry* (London: HMSO).

House of Lords Select Committee on the European Communities (1992), *Second Report: Review of the Common Fisheries Policy* (London: HMSO).

Leigh, M. (1983), *European Integration and the Common Fisheries Policy* (London: Croom Helm).

Symes, D. (1991), 'UK Demersal Fisheries and the North Sea: Problems in Renewable Resource Management', *Geography*, 76, 131–42.

Wijkman, P. M. (1982), 'Managing the Global Commons', *International Organization*, 36, 511–33.

Wise, M. (1984), *The Common Fisheries Policy of the European Community* (London: Methuen).

16 The UK and the International Environmental Agenda: Rio and After[1]

Michael Redclift

INTRODUCTION

Most of the chapters in this collection have examined, in some detail, the way that the British Government's environmental policy has evolved since 1988. Nowhere has this policy been faced with as many challenges as in the international arena. It was concern with possible global warming that prompted Mrs Thatcher's speeches on the environment to the Royal Society and to her Party Conference. But declarations and warnings need to be matched by deeds, and there are very few solid achievements in the realm of international environmental policy for which the British Government can take the credit.

Both Mrs Thatcher's administration and that of John Major have given verbal support to the idea of international agreements on the environment, but this has rarely been followed through with specific actions. The UK has done relatively little to strengthen its commitment to increased aid for developing countries. As I write, the controversy over the link between aid and defence contracts is revealing how many Conservative Ministers, and ex-Ministers, regard the aid policy as an extension of policy to boost British business. On the cancellation of the developing countries' debt, the British record is only marginally better. It should be remembered that although Mrs Thatcher's speech in 1988 came as something of a surprise, and by referring to 'global problems' she helped to legitimise the global environmental agenda in the UK, the references in her speech were exclusively to the evidence from 'hard science': ozone, acid rain and climate change. Biodiversity was not mentioned, and the Government's commitment was to 'sustainable economic development', rather than to the wider concerns of equity and empowerment signalled by 'sustainable development'. The commitment, at least in principle, of successive British Governments, has not been to tackle the underlying issues surrounding

283

international political economy, even in the limited sense advocated by the Dutch, the Canadians or the Scandinavians. The UK's response has been to *appear to support international action to address global problems, but to shy away from any tangible commitment to altering the global relationships which contribute to these problems.*

In our assessment, then, we need to ask three questions, which are closely related. First, how does UK policy measure up to the global environmental agenda? Second, *is* there an identifiable UK environmental policy at the international level? Third, what should such a policy look like, if it does not already exist? We will begin with the global agenda.

UNCED: THE ROAD TO RIO

Watching the media coverage of the UNCED at Rio de Janeiro in 1992 one could be excused for thinking that the principal environmental problems facing the globe were climate change and the loss of forests and, with them, biodiversity. There was very little coverage of the underlying problems which affect most human beings in their daily lives. *The Economist* complained that population growth was 'the issue that Rio forgot'. However, it was not the only forgotten issue, since there was little discussion, and less coverage, of a number of other important questions, notably trade, poverty and the continuing debt crisis in the South. With hindsight we may come to see UNCED as marking an important shift away from the development discourse of the 1970s and 1980s, towards a new concern with science and uncertainty, a concern that paralysed Northern governments by laying bare the contradictions of their development.

The environmental concerns of the 1990s have proved to be about the implications of profligacy, of plenty, rather than the limits to growth. The attention of the world came to rest not on the resources which made growth possible, but on the way that we dispose of the products of that growth, the policy we adopt towards 'sinks' (Brown, Adger and Turner, 1993b; Redclift 1994).

What was the 'state of the world' in August 1992 when the world's leaders met in Rio?

- More than two hundred million people were estimated to depend upon increasingly vulnerable forest resources.
- Eight hundred million people were affected by serious land degradation in dryland areas of the world.

- More than one billion people relied on fragile irrigation systems, plagued by problems of unreliable water supply and soil salinisation.
- Five hundred million people occupied degraded watersheds.
- Four hundred million people were vulnerable to the resulting down-stream siltation from these watersheds.
- An estimated one billion people lacked adequate, safe water for their own consumption.
- Almost two billion people suffered from either malnutrition or the lack of proper sanitation, *and each year thirteen and a half million children died as a result.* Two million of these children, under five years old, died from malaria.
- Over one billion of the world's population lived in cities where air pollution was below World Health Organisation (WHO) standards.
- Almost one billion people lived in cities where emissions of sulphur dioxide exceeded safe standards. (International Bank for Reconstruction and Development, 1992).

This was the very alarming context in which treaties were being prepared for signature at Rio. As I shall argue, the neglect of poverty and distributional issues generally has served to cast a long shadow over what was agreed at UNCED. It promises to be a shadow that will be difficult to cast off in the wake of the Rio agreements.

One of the things that distinguished the Rio conference from earlier meetings, including the deliberations of the World Commission for Environment and Development (Brundtland Commission), was the level of preparation, and the determination (on the part of at least some of the countries participating), to reach agreements. One of the main issues raised during the climate change negotiations, which preceeded UNCED, was how far 'responsibility' for emissions should be linked to 'obligations' that were binding on individual states. Before UNCED was convened there were meetings of various Working Groups and preparatory meetings (Prepcoms) intended to establish both the agenda and agreement about what was known and not known. At these meetings information was requested on a large number of environmental factors which made development unsustainable. It was information-gathering, rather than the formulation of policies to combat problems, which was the focus of attention.

These discussions gave rise to a number of problems, which were to emerge later at the conference itself (International Institute for Environment and Development, 1992). Ultimately, although there was general agreement that the North needed to cut its own emissions and to help the South cut its emissions (which was where the vexed question of 'additionality' came in), the Framework Climate Convention was vague and ineffective on both counts. It is also important to note that divisions within the South (and within the North) were important, and that some of the alliances of different interests were quite complex. Four divisive issues had already arisen.

First, the discussions on climate change were highly divisive. Most Southern countries wanted the focus to be on the transfer of cleaner technologies from the North, particularly for energy generation. This, in turn, raised the question of ways in which such transfers could be funded. There was also some discussion of the procedures for evaluating the costs of 'mitigating' the effects of climate change. Various countries from the South (the G77 group) argued that climate agreements should precede the forest agreement: that is, that the conference would need to tackle the causes of climate change first, rather than negotiate about the conservation of 'sinks', in the form of tropical forests. Such differences, of course, struck at the underlying orientations of the developed and developing countries: the former to prioritise climate change in defence of their development 'achievements', the latter to urge a different form of development on the North.

Second, the discussions surrounding biodiversity proved to be much more difficult than some observers had expected. There were problems of definition, as 'biodiversity' and 'biotechnology' were very closely related concepts. For the countries of the South with rich biodiversity, particularly wrapped up in tropical forests, there was a need to protect their resources, and their sovereignty, from the depredations of transnational companies located in the North.

In the view of most countries with large tropical forests, the search for biological materials involved the 'privatisation' of what were, in fact, common property resources. These sentiments were echoed by some (NGOs), and lobbyists in the North, for whom the ownership of nature was a political issue at least as important as the possession of global sinks.

Third, the preparatory meetings witnessed considerable discussion about the disposal of hazardous wastes, another issue which seemed to highlight the increasingly important 'sink' functions of the developing world. Sensitivities in this area were underlined by remarks attributed to a senior World Bank employee, Lawrence Summers, which implied that it was logical, on economic grounds, for richer countries to dispose of their wastes in the poorer countries, whose costs were lower. The discussions of hazardous wastes, prior to UNCED, focused on the feasibility of a global ban on their export to developing countries, and the increasingly urgent (with the dissolution of the Soviet Union) problem of disposing of radioactive wastes. Again, evidence was beginning to appear of mutually incompatible demands: from the North, that market forces should govern the disposal of wastes, from the South that the developed countries should begin by cleaning up their act.

The fourth set of issues to surface in the preparations for UNCED was arguably the most important. The absence of a real discussion of poverty struck at the very different perspectives of Northern and Southern governments. The G77 countries supported the idea that the Earth Summit, as it came to be called, should report on both *development* and the environment. It would be important, therefore, to consider ways of raising living standards in the developing world. They requested that population be left out of the Rio deliberations, particularly the linkages between population growth and the environment. The developed countries (G7) insisted that population was retained as an issue, reflecting their concern that rising living standards, implying increased levels of consumption in the larger countries of the South, posed the problem of accelerated global warming.

These four issues were interlinked in many ways: for example, there were heated discussions over responsibility for global environmental change, the North arguing that population was the driving force behind climate change, while the South insisted that 'overconsumption' in the North, at globally unsustainable levels, imperilled the survival of the world's poor. Linking these questions was the choice of mechanisms to affect a shift in the global imbalances. The transfer of cleaner technologies, for example, involved transferring financial resources to the developing countries through the Global Environment Facility (or GEF which was established in 1990), and there was a request for 'new sources of funding' to be identified before the conference convened. These issues, of the way technology was to be shared, and the financial mechanisms for ensuring that the South's role in global conservation was recognised, proved to be of enormous importance during, and in the aftermath of, Rio.

The process that led to Rio represented a development of that pioneered by the Brundtland Commission, with public fora being held in Bangkok, Mexico City, Cairo, Buenos Aires and Amsterdam, between February and May 1992. There was considerable participation from the NGOs, which sought to mobilise opinion around the production of national reports in advance of the meetings. These wide consultations between governments and NGOs led to the Global Forum at Rio, which paralleled the 'official' meeting of governments representatives. However, it should be noted that divisions within the ranks of NGOs became apparent as their public profile grew. O'Keefe, Middleton and Moyo (1993) make the point that the definition of what constituted an NGO was extraordinarily wide, including international businesses and, indeed, any organisation outside the public

sector. By the time UNCED was convened, the expectations for the conference far exceeded its mandate which was, as the International Institute for
Environment and Development declared, 'not very workable' (1992: 3).

THE UNCED DELIBERATIONS: CONVENTIONS AND A NEW AGENDA

In a meticulously documented and persuasive report, initially to the Dutch
Government, Krause, Bach and Kooney argued:

> Much of the current climate warming debate still proceeds along the
> narrow lines of conventional air pollution abatement policy. But climate
> stabilization is an entirely different challenge. The greenhouse effect is
> driven by a confluence of environmental impacts that have their source
> not only in the nature of human resource use, but also in the nature of
> the current international economic order. Climate stabilization *therefore
> requires a comprehensive turn around towards environmentally sound
> and socially equitable development – in short, an unprecedented North-
> South compact on sustainable development.* (Krause, Bach and Kooney,
> 1990: 1.7; emphasis added).

The failure of the Rio meeting was, ultimately, a failure to recognise this
relationship. Consequently, the omission of serious discussion about international debt, poverty and trade, as well as the failure to address population, served to undermine workable agreement on the issue of climate. The
International Institute for Environment and Development commented that
the problem existed even in the terms of reference for the conference:
'United Nations Resolution 44/228, December 1989 "did not set a very
workable mandate for the conference "' (1992: 3).

The discussion of climate was hamstrung by the very different perspectives of different groupings of countries. Large developing countries, like
China and India, refused to be thwarted in their efforts to develop, and to
improve the living standards of their vast, and very poor, populations. The
USA, and to a lesser extent the EC, sought to limit the effect of any international undertakings to curtail their economic growth, which were
viewed as politically inadmissible as far as their domestic audience was
concerned.

The governments subscribing to the Organisation of Petroleum
Exporting Countries clearly wanted to play a major part in managing the
world's dependence on fossil fuels, and did not favour transitional measures to move away from dependence upon them. The small island states

of the Pacific and Caribbean, in the front line of any projected sea-level rise, nonetheless possessed very little political influence. The countries of Sub-Saharan Africa were poorly represented at Rio and, even following the Nairobi Declaration (1990), viewed the climate negotiations as of relatively marginal interest. Within the South it was the so-called newly industrialising countries (NICs) such as Korea, Taiwan and Singapore which were most ambivalent about the climate negotiations. As O'Keefe, Middleton and Moyo (1993) argue, following the World Development Report (1992), Singapore and Hong Kong, between them, dispose of as much toxic wastes as the whole of Sub-Saharan Africa.

Initially the EC had sought to advance the idea of a unilateral carbon tax, as an instrument for enforced energy efficiency. However, this proposal failed when a 'conditionality' clause was inserted into the documentation requiring other developed countries to comply before the measure could be accepted. Japan and the USA both opposed the idea of a carbon tax, the latter government weakening the treaty further by refusing to endorse timetables for action and targets for greenhouse gas emissions.

The problems that accompany a comprehensive and workable treaty on climate have been discussed at length in the literature (see Grubb *et al.* 1993 for a definitive version of the agreements themselves). Among the most important are the following two points.

First, the Framework Climate Convention does provide a 'framework' for future action, if governments want to use it. They are not compelled to do so by the agreements entered into. The question remains: how can treaty commitments be made binding? As Grubb *et al.* argue:

> Thus it remains unclear how many aspects of the Climate Convention will operate in practice, and it can only be seen as an important first step on a very long road. The process of negotiating the Convention itself had an impact, both in educating many different countries, and in pushing concerned countries towards making and explaining commitments. (Grubb *et al.*, 1993: 72).

Second, one of the few sanctions that might be employed against climate convention 'free-riders' in the future, is some form of trade embargo, although this seems less likely now that an agreement on the GATT has been reached. If international environmental treaties use trade measures as disincentives to 'free-riding' then these measures will inevitably act as selective restrictions on imports, which (like the importation of goods produced 'unfairly') will contravene GATT. It is unlikely, then, that a climate agreement with effective powers of sanction can precede major new reforms in the GATT. As Paul Ekins has argued:

If the potential environmental benefits of free trade are to be realised, trading rules, such as those developed by GATT, must recognise that environmental externalities are in effect environmental subsidies (that are as economically distorting and unfair as any financial subsidy) and that environmental externalities are pervasive. (Ekins, 1993: 5).

Moving from the climate negotiations to the deliberations surrounding the declaration of Forest Principles, once again the division between the interests of North and South was apparent. The principle lying behind any kind of financial transfer to tropical forest countries was that preserving forests has a global environmental benefit which is not wholly captured by the host country. The industrialised countries could then be expected to transfer funds to poorer countries like Brazil or Indonesia, which possess large tropical forest reserves.

The background against which these negotiations took place was the much-discredited Tropical Forestry Action Plan, developed within the UN system. Opposition to this Plan from international NGOs and many developing countries was marked. Any agreement to manage tropical forests in the 'global' interest was likely to meet with objections from countries which, rightly, saw this as an infringement of their sovereignty. At the same time, financial compensation to such countries soon encounters problems of so-called 'additionality', which have dogged the role of the GEF. As Jordan has observed, without a real commitment to sustainable development there is little prospect of more funds being allocated to protecting the global environment:

At heart, the North is deeply opposed to providing new finance unless its self-interest is clearly guaranteed, whereas parts of the South simply lack the capability to cope effectively with a vast influx of new capital ... one suspects that, at root, there are very deep differences in the way that the North and the South perceive the need for, and likely role of additional financial transfers. The developed states seem to regard them as a price (to be minimized if at all possible) to be paid for enlisting the support of the South in tackling common problems. Meanwhile, the developing states seek to maximize transfers, arguing that they are a rightful and legitimate means to address the inequities in the operation of the world economy. (Jordan, 1994: 20–1).

The eventual outcome of the discussions of Forest Principles was a nonbinding and unclear set of propositions, which were opposed by leading developing countries, notably Malaysia and India. The G77 countries argued that they would only sign away part of their sovereignty in exchange for effective compensation for the revenue they would be forgoing. Eventually a compromise was brokered by the German Environment Minister, but unlike the Climate Change and Biodiversity conventions, no timetable was agreed, and the resulting document was not binding.

The outcome of the discussions surrounding biodiversity was rather different, in that in this case it was the industrialised world, principally the USA, which sought to advance its claim to acquire resources owned by another state. The developing countries wanted to control biodiversity as a means of protecting their stake in natural capital and the rent that this might provide. The North, on the other hand, wanted access to biodiversity for commercial profit, and particularly for the development of their research capacity in biotechnology. The USA felt the Treaty compromised its biotechnology industries, and refused to sign it.

Several options were put forward for breaking this inevitable deadlock. First, non-profit institutes in the North might contract for the use of wild genetic material, for commercial fees, or through sharing revenue. Second, private companies might support national scientific institutes in the South in return for exclusive rights to screen biological collections. Third, host governments might exploit resources, and foreign companies could be granted exploratory concessions by them.

The problem was: how can host communities acquire rights which translate into incentives to conserve biodiversity? This would either require large-scale international finance, dwarfing the modest proposed budget of the GEF or it would leave commercial operators free to exploit the market. In the event, after nearly four years of negotiation the Biodiversity Treaty was approved only in a much modified form. President Bush failed to ratify the Treaty, with the words, 'the American way of life is not up for negotiation' and although President Clinton ratified it in 1993 and the Treaty has now come into force, agreement on its long term funding has been delayed. The Treaty was described by Mostafa Tolba of the United Nations Environment Programme, as 'the minimum on which the international community can agree'.

IN THE WAKE OF RIO: INTERNATIONAL FINANCE AND POLITICAL DEVOLUTION

As Jordan has argued, the more specific the obligations to address global environmental issues, the more ambiguous and vague the commitments (Jordan, 1994). It is one thing to have developed a broad consensus of opinion on the need to tackle the issues and quite another to agree the formulae for funding the policy interventions that might ensue. We have seen how the commitments to part of the global agenda, referring to climate change and biodiversity losses in particular, principally reflects Northern concerns. The more fundamental questions for poor peoples' livelihoods in the foreseeable future, such as how to provide clean drinking water or to reverse land degradation, reflect aspects of sustainable development which were on the margins of the Rio discussions.

Agenda 21 was the document which paid most attention to sustainable development at the grass-roots level, and to the local management of resources. The UNCED Secretariat estimated that to implement *Agenda 21* would require *additional* aid of approximately US$125 billion worldwide *each year, from 1993 to 2000*. The estimates of the World Bank were somewhat more than half this figure. In the event, despite the rhetoric in evidence at Rio, only about US$2.5 billion was actually pledged by the richer countries. At a long pre-Rio negotiating session in New York, the French Government had proposed that the EU contribute US$3.8 billion to help pay for the implementation of *Agenda 21*. France's European partners were not easily convinced, and the scheme has effectively been abandoned.

Part of the problem in voting funds is that the institutions that manage them are scrutinised fairly closely. American environmentalists opposed the idea that the International Development Association, the 'soft'-loans element of the World Bank, should gain funds under its periodic budget replenishment (the tenth). Suspicion of the World Bank, or its subsidiaries, was widespread among NGOs, especially in the South. Many countries from among the G77 group wanted a 'Green Fund', financed by the North on an agreed formula, based on an international carbon tax. The two most important aspects of this proposal were the emphasis on sustainable development – rather than purely 'Global Environmental Change' questions – and the democratic management structure that was proposed. However, opposition to this idea from the G7 countries was vociferous and this idea, too, was dropped.

The GEF was formed before UNCED. It was to be administered by the World Bank (which provided the expertise for investment projects), the United Nations Environment Programme, which provided the Secretariat, and the United Nations Development Programme, for technical assistance and project identification. From the beginning, however, the GEF has been seen as the child of the World Bank.

The central principle behind the GEF's mandate was that of 'additionality'. The UNCED Secretariat suggested that the GEF could pay developing countries for the *extra cost* of policies to slow global warming, an idea which met opposition from the G7 countries. For many countries in the South there was little point in seeking to reach agreement on climate or biodiversity negotiations if the international economic order remained unchanged. For them the challenge was to redirect extensive funding towards the South, and to tackle essentially structural issues that impeded their sustainability.

The GEF itself reflected the position of the North. Voting rights in the GEF, like those in the World Bank, were heavily in favour of the developed countries. The staff of the GEF were recruited from the World Bank. Most importantly, however, the terms of reference for the GEF's work were essentially those agreed by the donor countries.

The remit of the GEF was exclusively with 'transnational' environmental problems, such as transboundary pollution, and the loss of biodiversity, rather than the 'bread and butter' issues which concerned most of the developing world. In addition, the developing countries expressed the view that progress depended upon the cancellation of their debts. Without the cancellation of the debt they saw the demand for environmental good practice as, in effect, a form of conditionality, which would be used to ensure their compliance with the wishes of the donor countries.

As Jordan demonstrates, the vexed question of environmental 'additionality' was inherited from the negotiations over the Montreal Protocol, when countries like China and India acceded to the agreement because they saw the prospect of new funds (Interim Multilateral Ozone Fund), and financial inducements to clean technology transfer (Jordan, 1994). In practice, however, the developed states favoured the Bretton Woods institutions (World Bank and the IMF) because they were deemed to be more efficient, and their work of more technical merit. They were also more accountable to their paymasters within the G7 community of nations. The work of the GEF would essentially bolster some of the activities with which the World Bank was already engaged, attaching a price to existing or augmented project funding. The wider aspirations of *Agenda 21*, and the search for sustainable development, could be politely forgotten.

Almost before the last government representatives had left Rio it became clear that the ambiguous wording of the agreements, and the exemptions granted to individual states, would make it difficult to follow up even the rather 'tighter' Framework Convention on Climate Change and the Biodiversity Convention. Their management was entrusted to the GEF, but only on an interim basis. The existence of concrete funds with which to meet the 'additionality' provisions, was soon put in jeopardy. As Jordan comments, 'at heart, the North is deeply opposed to providing new finances unless its self-interest is clearly guaranteed' (Jordan, 1994: 20).

Even the sums of money which remain contested, however, are so modest as to make no demonstrable difference to the problems they are designed to address. Even a sympathetic view of the GEF's work elicits the conclusion that without substantial debt relief, for example, little can be achieved within the terms agreed at Rio (Brown, Adger and Turner, 1993b).

In June 1994 the GEF reaches the end of its three year pilot stage. A 'replenishment' of about US$2 billion has been proposed for its continuation. The future of the GEF was discussed at a meeting in Cartagena, Colombia in December 1993. An evaluation of the GEF's work reflected opposition from some G77 governments, and the continued doubts of environmental NGOs. The dilemma faced by the GEF is that it needs to reassure the North to retain any funding and credibility, without entirely losing the confidence of the G77 countries, whose consent to its activities is essential.

Against the wishes of the World Bank, the evaluation of the GEF suggested that all finance for new projects should be suspended until new guidelines were in place. The evaluation stated that the definition of 'incremental cost', to be met by the GEF, as 'the additional cost of protecting the environment of a country, over and above its domestic benefits', was insufficiently precise. Even more contentiously the idea has been put forward that GEF projects should not be initiated by the Bank, the UN Development Programme or the UN Environment Programme, but at the regional and national levels of developing countries.

In addition, it was suggested that regional development banks, other UN agencies and environmental groups should be able to compete for the GEF's funds, and that NGOs and developing country governments should be more involved in the GEF's decisions. None of these suggestions is popular with the senior staff of the World Bank. The result has been a stalemate, the World Bank arguing that progress has been made in devising mechanisms to manage the new facility, while the developing country representatives maintain that they are being offered less money and no effective managerial control. As one Latin American representative put it: 'offering money in exchange for control over an organism in which the North and the South should be partners shows us that there are still walls more difficult to tear down than the Berlin wall' (Raghavan, 1993: 4).

THE UK AND THE INTERNATIONAL ENVIRONMENTAL AGENDA

The questions posed at the beginning of this chapter need to be addressed in the light of the developments at the international level which have been documented. To what extent does the UK's policy at the international level measure up to the challenges of UNCED?. Has the UK influenced the policy agenda in a positive fashion? What measures could the UK take to advance the wider interests of the global community, and would such measures be at the cost of its own national interests? We can begin with an assessment of the UK response to UNCED. In their analysis of this response Brown and Jordan argue that: 'The evidence, to date, suggests a cautious and nationalistic response from the UK, and a lack of willing cooperation with other European states over timetabling, targets and burden-sharing arrangements' (Brown and Jordan, 1993: 88). They reach this verdict after giving some attention to a number of important potential outcomes of the UNCED agreements. First, the UK, as a signatory to the United Nations Framework Convention on Climate Change, has undertaken to communicate detailed information to the Conference of the Parties on the measures taken to limit emissions, together with information on the level of emissions, with the objective of returning emission levels to those of 1990 by the year 2000. This would include a national inventory of sources and sinks, of all GHGs, as well as a detailed description of the policies and measures that it proposes to undertake (Brown and Jordan, 1993: 89).

The state of the 'science' on anthropogenic emissions is itself somewhat confused, partly because of the difficulty in distinguishing clearly between emissions (and sinks) from natural systems, and those which result directly from human activities. There are also largely unresolved issues surrounding the relative importance of different GHGs, and the time frame in which their impact is assessed (Adger and Brown, 1993). However, while these difficulties serve to complicate the picture, they do not obscure the fact that the UK's modest achievements in slowing down emissions are attributable to the recession in the British economy, rather than to far-sighted policy measures.

A number of policy changes within the UK have made the achievement of emission targets more difficult. First, the British Government's deregulation policies have hindered, rather than helped, meeting these targets. As Brown and Jordan point out, 'rolling back the state ... [has] lessened the leverage that the Government can bring to bear on the climate change issue' (Brown and Jordan, 1993: 90). More attention needs to be paid to the impact of a broad swathe of policy measures, including the privatisation of the railways, and the changes in the energy supply industries.

Second, there is very little evidence that the Government is pursuing a policy of encouraging energy efficiency, or of facilitating even a very gradual transition towards more dependence on renewable sources of energy. The initiative for energy efficiency has been laid at the door of individual households, rather than at the door of central or local government. Imaginative measures to combine heat and power generation, and to manage the production of energy through incentives on the demand side, such as those pioneered in parts of California, have been given no effective support.

Third, the current emphasis within UK Government circles has been on the need to reach agreement, rather than to advance policies which could secure this agreement through underwriting policy shifts. For example, Britain was also a signatory to the Biodiversity Convention, as we have seen, but little has been done to protect its own biodiversity, as other chapters in this volume attest. Many other policies affecting the environment and land-use, such as the proposed expansion in motorway construction (now under review), and the continued support for the EU's CAP, are likely to damage, rather than conserve, biodiversity. Policies such as 'set-aside', where they are employed, have been prompted by the need to keep the support of the farming lobby, rather than by clear environmental conservation objectives.

It is curious that the British Government has given support to public fora which emphasise, in terms of critical media coverage, the weaknesses of its international environmental policies, rather than the strengths. On the other hand, the Government has funded research in the social sciences, to the tune of £20 million, over a ten-year period to examine the causes and consequences of global environmental change, under the aegis of the ESRC. It has also put approximately £2 million a year into the much-vaunted 'Darwin Initiative', to be administered by the DoE, with advice from the Natural Environment Research Council. However, such research programmes serve to point out what can and should be done to mitigate and avoid global environmental changes, and reveal how much more needs to be done in the UK.

Also the Government has supported a series of conferences, emulating the Earth Summit, at which both Government and NGO representatives will debate the issues surrounding global environmental change. The first of these was the 'Partnership For Change' conference held in Manchester in September 1993. Other meetings have accompanied the launch of the consultation document 'UK Strategy For Sustainable Development', to be submitted to the Sustainable Development Commission. The Summary Report which followed this document, entitled *Sustainable Development: the UK Strategy*, was published in January 1994.

The publication of these strategy documents might appear to augur a new phase, under which the British Government is preparing to take the initiative. According to *Climate Change: The UK Programme*, the report dealing with Britain's obligations under the Framework Convention, the UK is the first country to 'demonstrate how it will [return emissions of GHGS to 1990 levels by 2000]' (DoE 1994b). At face value, such a claim is impressive. Another accompanying document, *Biodiversity: The UK Action Plan* (DoE 1994a) makes similar claims to novelty and innovation. Finally, *Sustainable Forestry: The UK Programme* is described, pending the planned Forestry Review (due to report in 1995) as 'a comprehensive statement of the Government's present policies' (DoE 1994c).

The publication of these reports does not invalidate the criticisms directed at the earlier consultation papers, on which the reports are based. There are several underlying problems with each of them.

First, the British Government response does not establish ground rules for sustainability planning, such as those put forward by Jacobs (1991). It simply advocates 'business-as-usual' planning, without acknowledging the very real constraints on future economic growth itself represented by the use of resources.

Second, the commitment to deregulation, including the deregulation of international trade under the reformed GATT arrangements, provides no protection for environments, and no means of ensuring that environmental costs are fully represented. As Daly and Goodland (1992) have demonstrated, the implications of our trading relations are fundamental to successful sustainable development, and the gains from free trade are likely to be more than offset by the social and environmental costs.

Third, as Friends of The Earth (UK) claim, in their response to the early documentation provided by the Government, nowhere is there recognition of *fundamental principles* for protecting the environment, such as the precautionary principle, as a means of setting workable targets (Friends of the Earth, 1993). Similarly, the Government's strategy needs to use economic instruments as tools of policy, rather than 'as a means of setting objectives of policy' (Friends of the Earth, 1993: 89). The action advocated under *Agenda 21* and *Towards Sustainability*, the EU's policy statement, remains very modest. The evidence is that the British Government is unwilling to ensure even these limited objectives the success that they deserve.

TAKING UP THE CHALLENGE: WHAT SHOULD BE DONE?

If we conclude that the British response to the international environmental agenda has been largely fashioned by self-interest, without giving serious consideration to measures which incorporate more sustainable development into existing policies, then the obvious corollary is: *what should be done?*

There are a number of responses to this question, most of them building on existing environmental practices, or seeking to develop new forms of collaboration within civil society. First, there is little point in considering global environmental changes as unrelated to other global processes, such as the liberalisation of trade and the penetration of the media.

Responding positively to the international agenda on the environment means establishing the connections between environmental benefits and other social and economic benefits and costs. The recent White Paper on Science and Technology manifestly fails to do this, by adding on 'quality of life' to 'wealth creation' almost as an afterthought. In fact, we need to begin to *define wealth creation in terms of the quality of life.*

At the international level this inevitably takes us back to some of the discussions prior to the UNCED meeting, to consider the role, for example, that 'debt forgiveness' might play in channelling resources towards sustainable development projects in the South. The UK has enormous expertise, unrivalled for a country of its size, in the technical and managerial aspects of project management and appraisal in developing countries. What is required is a commitment to exercise these skills on behalf of the developing nations.

Second, the global environmental policy agenda cannot be fashioned around a belief in impartial 'scientific' advice. It is true that some of the research institutions of science and technology perceive short-term benefits in the uncertainty surrounding global warming. However, it may be more difficult to reach agreement when the facts are 'uncertain', and heavily contested by different lobbies or 'stakeholders'. The answer is not to call for an end to uncertainty – or to wait until such an end is in sight – but to take preventative measures now, most of which are beneficial in the short term. To be successful in such a strategy means enlisting the public, in a way never contemplated by governments outside wartime.

Third, the broadening of the debate about global change, and the increased involvement of the public, rests upon a process of democratisation and a new definition of participatory citizenship. It may well be at the level of local government that this proves to be most important. It can certainly be argued that *Agenda 21* has proved most useful as the basis for wide-ranging inquiry, on the part of local authorities themselves, into the best methods of increasing local-level sustainability.

Much closer links are needed between local voluntary groups, properly funded statutory bodies and local government. To develop these links would not only demand a devolution of power and resources downwards, towards elected representatives and local organisations, but also a willingness to undertake a dialogue with several 'sides' to it. The outlook of the present British Government seems to be in favour of more local 'managerialism', less accountability and policy solutions based on fiscal economies.

Fourth, perhaps the most important level at which the British Government could make a contribution, is the one for which it shows least enthusiasm: that of the EU. Liberatore (1993) has shown how the UK has failed to rise to the European challenge. She writes that:

> The European Community provides the unique example of a supranational setting since the EC legislation supersedes national legislation and an EC institution (the Commission) has regulatory competences...the EC willingness to play a leading role is linked to the need to formulate an agreed position for the Community and its Member States. (Liberatore, 1993: 15–16).

The strategy of the EU, in response to the challenges of UNCED, consists of several parts, each of which requires British endorsement.

1. The EU can act in a leadership role internationally, both because of its size and the wealth of its constituent states, and because it has long recognised the need to curb the excesses of economic growth. The opportunity exists to play an enhanced collective role on the world stage. The move towards an 'internal market' within the EU is expressly linked to that of enhanced sustainability. The British Government should welcome this commitment, rather than resenting control from Brussels.
2. The fulfilment of environmental, as well as social and economic goals, presents real opportunities for investment in cleaner technologies, and for environmental protection measures which enhance economic competitiveness. Measures that form part of a no-regrets strategy (such as the establishment of more ambitious targets for the stabilisation of emissions), measures aimed at increasing energy efficiency and the carbon dioxide/energy tax could form the basis of a new global policy initiative. Instead, the UK expressed scepticism, even before UNCED, about the effects of such measures. In addition it partly shared the scepticism of the USA towards the biodiversity convention, and the need to increase development aid significantly (Liberatore, 1993: 21).
3. In principle, at least, *Towards Sustainability*, the EU's programme for sustainable development, and Fifth Environmental Action Programme, which was published in 1992, has a number of relatively innovative ingredients. It calls for preventative measures to deal with

environmental problems, and identifies the failure of *post hoc* environmental management. It also seeks to involve all sectors of society in the design of solutions to our current environmental problems.

These problems clearly extend beyond the nation state and beyond regional groups like the EU itself. Recognition that this involves responsibilities for governments, however, can be seen as an enormous policy opportunity: to put flesh on the bones of the Environmental Action Plan would be a significant new step.

CONCLUSION

I began this chapter by arguing that Britain had played little more than a cosmetic role in promoting the Earth Summit, the event being seen as an opportunity to appear more proactive than was really the case. Events since 1992 confirm that we have failed to enter into the spirit of the new international agreements, however flawed they may be, and still seek to benefit from the confusions surrounding them, rather than working towards a more sustainable world order, in which the UK might play a leading role.

Note

1.   I would like to express my thanks to Andrew Jordan, Kate Brown and Neil Adger, of CSERGE (UEA) for their helpful and constructive comments on an earlier draft of this chapter.

References

Adger, N. and Brown, K. (1993), 'A UK, Greenhouse Gas Inventory: On Estimating Anthropogenic and Natural Sources and Sinks', *Ambio*, 22, 509–17.
Brown, K., Adger, N. and Turner, K. (1993a), 'Forests For International Offsets: Economic and Political Issues of Carbon Sequestration?', *CSERGE Working Paper*, GEC 93–115.
Brown, K., Adger, N. and Turner, K. (1993b), 'Global Environmental Change and Mechanisms for North-South Resource Transfers', *Journal of International Development*, 5, 571–89.
Brown, K. and Jordan, A. (1993), 'The UK Government's Response to UNCED', *Pacific and Asian Journal of Energy*, 3, 87–99.

Daly, H. and Goodland, R. (1992), *An Ecological-Economic Assessment of Deregulation of International Commerce Under GATT* (Washington, DC: Environment Department, World Bank).

DoE (1994a), *Biodiversity: The UK Action Plan* (London: HMSO).

DoE (1994b), *Climate Change: The UK Programme* (London: HMSO).

DoE (1994c), *Sustainable Forestry: The UK Programme* (London: HMSO).

Ekins, P. (1993), *Trading off the Future* (London: The New Economics Foundation).

Friends of the Earth (1993), *Nothing Ventured, Nothing Gained* (London: Friends of the Earth).

Grubb, M., Koch, M. Munson, A. Sullivan, F. and Thomson, K. (1993), *The Earth Summit Agreements* (London: Earthscan).

International Bank for Reconstruction and Development (1992), *World Development Report* (Washington, DC: International Bank for Reconstruction and Development).

International Institute for Environment and Development (1992), 'Rio: The Lessons Learned', *Perspectives*, No. 9, Issue title.

Jacobs, M. (1991), *The Green Economy* (London: Pluto Press).

Jordan, A. (1994), 'Financing the UNCED Agenda: The Vexed Question of Additionality', *Environment*, March: 16–27.

Krause, F., Bach, W. and Kooney, J. (1990), *Energy Policy in the Greenhouse* (London: Earthscan).

Liberatore, A. (1993), *Beyond the Earth Summit*, European University Working Paper 93/5.

O'Keefe, P., Middleton, N. and Moyo, S. (1993), *Tears of the Crocodile* (London: Pluto Press).

Raghavan, C. (1993), 'GEF Meet Ends in Disagreement', *Third World Resurgence*, 41, 17–29.

Redclift, M. R. (1994), 'The Other Side of Consumption: Towards a Theory of Sinks', *MS*.

Index

acid rain   7, 68, 144, 173, 176,
    189–209
Agriculture Act (1986)   8, 240
air pollution   14, 24, 31, 60, 88,
    96–7, 168, 193
Alkali Acts   60
Alkali Inspectorate   48
    see also Her Majesty's Inspectorate
        of Pollution
Areas of Outstanding Natural Beauty
    (**AONBs**)   76
Areas of Special Scientific Interest
    (**ASSIs**)   89, 91
    see also Special Sites of Scientific
        Interest
Atkins, Robert   92, 95–6
Audit Commission   152, 154

bathing water   22, 24, 28
    see also European Commission
        Directives
Belgium   34, 205
biodiversity   76, 176, 237, 283–4,
    296
    see also Rio Earth Summit
Biodiversity: the UK Action Plan
    (1994)   246, 250, 254, 297
birds   22, 24
    see also European Commission
        Directives
Black Report   212
Bottomley, Peter   86, 91, 98
Brazil   290
British Board of Agreement (**BBA**)
    168–9
British Coal   198
British Gas   150
Brundtland Commission and Report
    13, 246, 285, 287
Brussels   113, 256, 278, 299
Bush, George   16, 291

Campaign for Lead-Free Air
    (**CLEAR**)   18
    see also lead-free petrol
Campaign for Nuclear Disarmament
    (**CND**)   218
Canada   276, 284
Cap de la Hague   217
carbon dioxide emissions   6, 144,
    159, 162–3
    see also global warming, climate
        change
carbon energy tax   5, 123–4, 126–8,
    141, 149
    see also European Carbon Energy
        Tax
Chernobyl   22, 251
China   288, 293
Clarke, Kenneth,   133, 138
Clean Air Act (1968)   30
climate change   124–8, 136, 147,
    196, 284, 288
    Inter-governmental panel on
        (**IPCC**)   124–5, 127, 147
    see also global warming
Climate Change Convention   *see* Rio
        Earth Summit
Clinton, Bill   14, 291
coal   129–30, 139, 157, 197–8, 203
    see also fuel, mining
combined heat and power (**CHP**)
    151, 155
Common Agricultural Policy (**CAP**)
    9, 240, 296
Common Fisheries Policy (**CFP**)   9,
    263–82
    'days at sea'   273–5, 281
    see also fishing
Confederation of British Industry
    (**CBI**)   43, 48, 53, 78
Conservative Party   27, 141
    manifesto (1992)   38
    Euro sceptics   22, 27, 35

Control of Pollution Act (1974)　44
Council for European Municipalities
　　and Regions **(CEMR)**
　　Environment Committee
　　(previously called the
　　Environment Dialogue Group
　　(EDG) 106–7, 113–14, 120
Council for Nature Conservation and
　　the Countryside **(CNCC)** 4,
　　91–2, 99
Council for the Protection of Rural
　　England **(CPRE)** 53, 243, 254
Council of Europe Convention on
　　Civil Liability 79
Country Landowners' Association
　　(CLA) 53
Countryside Act (1968) 237
Countryside Commission 50, 240,
　　244–5
Countryside Council for Wales
　　(CCW) 39, 245, 250
'critical natural capital' 75–6
cross-border pollution 27, 31
Cumbria County Council 229

Denmark 29, 34, 73, 205, 265, 270,
　　276
Department of Energy 144
Department of Environment **(DoE)**
　　15–17, 23, 42, 49, 52, 106, 112,
　　119, 125, 145, 147, 149, 190–1,
　　220, 253–4
　　Environment Agency 38–56
Department of Environment in
　　Northern Ireland **(DoE NI)** *see*
　　under Northern Ireland
Department of Trade and Industry
　　(DTI) 147, 151, 191–2
Department of Transport **(DoT)** 16,
　　32, 173–88, 247, 254
di Meana, Carlo Ripa 22, 25–7
Directorate General for the
　　Environment **(DGXI)** 21–30,
　　32–4, 105–7, 109–11, 120
Dounreay 221
Drinking Water Inspectorate 92

'ecological buffers' 75

electricity supply industry 150–1,
　　156–7, 189, 191–2, 197–203
energy conservation 5–6, 126, 135,
　　144–58, 161–4
Energy Efficiency Office **(EEO)**
　　144–5, 157, 169
Energy Management Assistance
　　Scheme **(EMAS)** 136, 149
Energy Saving Trust **(EST)** 5, 135,
　　150–1, 155–7
Energy Select Committee 147–8,
　　154
England 49, 86–7, 90, 119, 124–5,
　　131, 157, 179, 195, 213, 244–6,
　　249, 278
English Nature **(EN)** 50, 52–3,
　　244–5, 248, 250
　　see also Nature Conservancy
　　Council
Environment Agency **(EA)** 1, 3,
　　38–56
Environment Bill (1994–5) 54–5
Environment Committee (House of
　　Commons) 40, 43–4, 86, 93,
　　97, 152, 154, 156–7
Environmental Action Programmes
　　(of EU) 10, 21, 33, 105, 204
Environmental Health Officers 111,
　　113
environmental impact assessment
　　(EIA) 60
　　see also European Commission
　　Directives
Environmental Protection Act 1990
　　(EPA) 1, 30–1, 42, 44–5, 60,
　　78, 195, 244
Environmentally Sensitive Areas
　　(ESAs) 8, 240, 247–51, 254
European Commission Directives
　　Air Framework (1984) 200
　　Bathing water (1976) 22, 109
　　Birds (1979) 22, 237
　　Construction Products 162, 170
　　Drinking Water (1980) 22, 79
　　Environmental Impact Assessment
　　　(1980) 23–6, 252
　　Gas Oil 206
　　Habitats (1993) 76

Large Combustion Plant (1988) 7, 189–90, 194–8, 204–5
Lead Free Petrol (1985) 18
Toxic Waste (1978) 20
Urban Wastewater Treatment 77
European Carbon Energy Tax 5, 138–9, 141, 149
European Communities Act (1972) 23
European Council of Ministers 27, 30–1
European Court of Auditors 104
European Court of Justice (ECJ) 21, 23–4, 27, 30–1, 35, 252, 274–5
European Elections
1989 26, 251
1994 34
European Environment Agency 31
European Regional Development Fund (ERDF) 110
European Union's Fifth Environmental Action Programme 10, 105–6, 115, 204
European Union Financial Instrument for the Environment (LIFE) 110, 112,
Exxon Valdez 251

fishing 9, 263–82
'Save Britain's Fish' (SBF) campaign 279–80
see also Common Fisheries Policy
flue gas desulphurisation (FGD) 194–5, 197–9, 201
Foreign Office 27
Forestry Commission 8, 50, 240–1, 245
France 20, 34, 125, 205, 211, 223, 271, 292
Friends of the Earth 11, 48, 97–8, 219, 297

gas (natural) 149–51, 155–7, 193
General Agreement on Tariffs and Trade (GATT) 179, 289–90, 297

Germany 3–4, 34–5, 57–8, 62–3, 68–9, 71, 73–4, 205, 223, 233, 291
Global 2000 Report 13
Global Environment Facility (GEF) 287, 292–4
global warming ('greenhouse effect') 5, 22, 68, 97, 123–44
see also climate change
Greece 34, 175
Green Parties
in EU 22
in Germany 68
in UK 26, 251
Green Alliance 11
'greenhouse effect' *see* global warming
greenhouse gases (GHGs) 123–9, 176, 179–80, 212, 295
Greenpeace 63, 95, 98, 213, 215, 220, 232
Gummer, John 39, 55, 224

Hague Declaration (1989) 125
Health and Safety Executive 30
Her Majesty's Inspectorate of Pollution (HMIP) 3, 14, 31, 38–40, 42, 46, 48–54, 92, 192, 200–1, 213
Heseltine, Michael 39, 43, 49, 128, 251
Home Energy Efficiency Scheme (HEES) 149
Hong Kong 289
Houghton, Dr John 125
House of Commons Select Committee on the Environment 40, 43–4, 48–9, 54
House of Lords 79
House Builders' Federation 166–7
Howard, Michael 39, 49, 54
Hurd, Douglas 27

India 288, 290, 293
Indonesia 290
Inter-departmental Group on Environmental Economics (IGEE) 126–7

Intergovernmental Conference (1996)
34
Intergovernmental Panel on Climate
Change (**IPCC**)　124–5, 127,
147, 204–5
International Monetary Fund (**IMF**)
23
Institute for European Environmental
Policy (**IEEP**)　3
Institute of Water and Environmental
Management (**IWEM**)　53–4
✱ integrated pollution control (**IPC**)
42, 47, 52, 88
✱ integrated pollution prevention control
(**IPPC**)　204–5
✱ Integration Pollution Regulation
(**IPR**)　200–1
International Atomic Energy
Authority (**IAEA**)　219
Ireland, Republic of　35, 86, 95, 265,
270, 276

Japan　62, 154, 228, 233, 289
JIMBOB places　241, 248

Korea　289
Kramer, Dr Ludwig　21, 24, 29

Labour Party　39
La Hague　233
Lamont, Norman　132–4, 138
Lancashire County Council　109
lead-free petrol　15, 18, 22, 31, 177,
182
see also European Commission
Directives, **CLEAR**
local authorities　4–5, 44–5, 101–22
Association of Metropolitan
Authorities　103, 112
International Union of Local
Authorities　106
Local Government International
Bureau (**LGIB**)　113–14
Local Government Management
Board (**LGMB**)　103, 106

Maastricht, Treaty of　14, 21, 26–7,
32–4, 69–70, 104, 138, 141
Major, John　27, 38–40, 124, 127–8

Malaysia　290
methane emissions　130, 139–40
Ministry of Agriculture, Fisheries and
Food (**MAFF**)　8, 48–9, 53,
213, 252, 254
Montreal Protocol on CFCs　35, 125,
183, 293

Namibia　276
National Audit Office (**NAO**)　49
National Farmers' Union (**NFU**)　53
National Federation of Anglers　53
National Federation of Fishermen's
Organisations (**NFFO**)　271,
273–5
National House Builders' Council
(**NHBC**)　166–7
National Parks　52, 237, 252
National Rivers Authority (**NRA**)　3,
8, 14, 25, 30, 38–40, 42, 46–55,
78, 92, 245, 247
Nature Conservancy Council (**NCC**)
8, 52, 90–1, 239, 242, 244, 246,
251
Netherlands　33, 35, 73–4, 125, 205,
265, 284
nitrate pollution　29
nitrous oxide emissions　29, 130,
139–40, 144, 176, 193, 195, 199
Nitrous Oxide Protocol (1988)
204
North Sea
Conference (1987)　77
gas　86
Ministerial Declaration on (1990)
69, 73
oil　203, 206
pollution of　68
Northern Ireland　4, 85–100, 195
Department of Economic
Development (**DED**)　94
Department of Environment, NI
(**DoE NI**)　4, 89–90, 92–3,
97–9
Rossi Report　86–9, 92–3, 97
Norway　125, 265, 269, 276
Nuclear Industry Radioactive Waste
Executive (**NIREX**)　8, 210,
217, 219, 221, 224, 227–33

nuclear power   126, 129, 155, 157, 210–36
  disposal of waste   7–8, 210–236
  British Nuclear Fuels Ltd **(BNFL)** 210–13, 223–7, 232–3
  Sellafield   211–2, 221–2, 226, 230–3
  Thermal Oxide Processing Plant **(THORP)**   210–15, 218, 220–8, 231–3

oil   14, 149, 206
organic farming   96
Organisation for Economic Cooperation and Development **(OECD)**   13
Organisation of Petroleum Exporting Countries **(OPEC)**   288
Oxleas Wood   24, 248, 252
ozone layer depletion   14, 22, 35, 59, 75, 125, 176

Patten, Chris   40, 244, 251
Plantlife   245
Poole Council   247, 252
Portugal   276–7
post-materialism   62
precautionary principle   3–4, 57–84, 147
privatisation (of utilities)   1, 53, 92, 150

Ramsar Convention   249, 251
Ridley, Nicholas   40, 244, 253
Rio Earth Summit (United Nations Conference on Environment and Development – UNCED) 1992   1–2, 5–6, 9–10, 13, 15, 57, 67, 69, 124–5, 127–8, 144, 183–4, 219, 246, 255, 283–301
  Agenda 21   13, 219, 292, 297–8
  Conventions on Climate Change and Biodiversity   5–6, 13, 67, 124, 127–8, 183–4, 246, 289, 291, 293, 295–7
  Declaration of Principles   13, 57, 67, 69
  Forest Principles   290
road building   *see* transport

Rossi Report 1990   *see under* Northern Ireland
Royal Commission on Environmental Pollution   41, 50, 52, 71, 220
Royal Society for Nature Conservation **(RSNC)**   97
Royal Society for the Protection of Birds **(RSPB)**   90, 97, 243, 245
Russia   125, 220, 286

Scotland   39, 49, 90, 119, 195, 213, 244, 251, 273
Scottish Natural Heritage **(SNH)**   39, 244, 250
Scottish Office   253
Sea Fish (Conservation Act) (1992)   9, 273–5
Seveso   19
Singapore   289
Single European Act   21, 101, 104–5
Soviet Union   *see* Russia
Spain   271, 276–7, 279–80
Special Areas of Conservation **(SACs)**   76
Special Sites of Scientific Interest **(SSSIs)** (ASSIs in N. Ireland)   8, 76, 89, 91, 237–8, 242, 246–9, 251
'strict liability'   66, 78–9
subsidiarity   27–8, 46, 104
sulphur dioxide   7, 89, 97, 144, 176, 195, 197, 199
sustainability   10, 75, 85, 204, 264, 283, 298
Sustainable Development: The UK Strategy (1994)   1, 70, 144, 296
Sustainable Forestry: The UK Programme (1994)   247, 297
Sweden   211
Switzerland   252

Taiwan   289
Thatcher, Margaret   1, 2, 10, 39, 124–5, 190, 196, 237, 239, 247, 251, 283
This Common Inheritance (White Paper) (1990)   1; 32, 39–40, 43, 71, 124, 126–8, 137, 196, 244

Third Force Organisations   *see under* wildlife
THORP   *see under* nuclear power
Tolba, Mostafa   291
Toronto Agreement   125
toxic waste   20
transport   6–7, 24, 97, 127, 173–88, 252
  road fuel duty   131, 133–4, 140, 149
  roads   24–5, 27, 134, 140, 155
Transport 2000   98
Treaty of Accession (1972)   265, 276
Treaty of Rome (1956)   104, 279
Trippier, David   40
Twyford Down   24–5, 252

UNCED   *see* Rio Earth Summit
UN Conference on the Human Environment, Stockholm (1992)   12
UN Environment Programme (**UNEP**)   291–2
Urban Environment Expert Group Committee (or Urban Environment) (**UEG**)   106–7
USA   15, 62, 125, 175, 211, 217, 220, 223, 233, 276, 289, 291

VAT on domestic fuel   15, 123–4, 131–4, 140, 149
Vorsage   67–8, 71, 74

Waldegrave, William   40
Wales   39, 49, 87, 90, 119, 131, 157, 195, 213, 244–6, 249, 253, 278
Walker, Peter   144
Waste Regulation Authorities   42
water
  authorities   31
  privatisation   25, 53, 92
  regulation   8
Water Research Centre (**WRC**)   77
Water Acts 1989 and 1991   30, 42, 78, 245
Welsh Office   52, 253
White Paper 1990   *see This Common Inheritance*
wildlife   8–9, 90–1, 98, 237–62
  Third Force Organisations (**TFOs**)   255–8
Wildlife and Countryside Act (**W&CA**) 1981   8, 237, 239, 247
  Amendment Act   238, 240
Windscale Inquiry   220
World Bank for Nature   292–4
World Wildlife Fund   243, 245